THE ART OF GERM WARFARE

"LEARNING TO READ THE BATTLEFIELD"

APOSTLE SANDY LOVE

Issachar's Oil Publishing House

ISBN: 978-1-957802-15-2

Issachar's Oil Publishing House

THE AUTHOR'S DESK

The study of germ warfare has always fascinated me, and I have dedicated two decades to understanding this critical aspect of modern warfare. While I have previously taught this subject to younger audiences, it is essential to tailor the discussion for adults to truly grasp its significance. As a former military member and the spouse of a soldier for over twenty years, I bring a unique perspective to this topic—one that is often inaccessible to civilians.

Although Christians view us as soldiers in the Army of the Lord, we often lack the knowledge to navigate the complexities of our daily struggles. By exploring germ warfare, we can gain insight into our challenges and become better equipped to defend our faith and values.

Considering how to prepare for battle even before stepping onto the battlefield is intriguing. "The Art of Germ Warfare" offers a unique perspective, serving as an upgrade element or a tool to enhance one's skills on the battlefield of prayer. Unfortunately, reducing war casualties is often overlooked, and I hope this book will draw more attention to this critical issue.

"The Art of Germ Warfare" is a practical guide that can equip you with the knowledge and skills needed for the battlefield. I encourage you to absorb its wisdom and use it as a reference in your future battles.

1

THORNS BIRTH REVELATION

Paul, a significant figure in the early Christian church, humbly referred to himself as the least of the apostles despite his substantial contributions to spreading Christianity. It is said that to prevent Paul from becoming prideful or overly confident, God allowed him to experience a "thorn," a physical ailment. This thorn served as a reminder of his human limitations and the importance of humility in the face of divine grace.

REFLECTING on Paul's experience offers a broader lesson about the nature of spiritual journeys and the challenges accompanying a prophetic calling. Just as Paul faced his thorn, individuals called to speak truth to power or deliver unpopular messages often encounter their challenges. This rejection can be a significant burden for those on the path of prophecy.

HOWEVER, the concept of thorns in the context of spiritual growth extends beyond mere rejection. It illustrates that everyone, regardless of their specific calling, will face obstacles and challenges along their journey. These thorns may take different forms for different people, but they share a common purpose: to keep us grounded, humble, and reliant on the divine for strength and guidance.

Like Paul, those who embrace their prophetic calling must be prepared to bear a heavy cost for their beliefs. This cost may manifest as personal sacrifice, persecution, or hardship. Yet, through it all, the thorns we encounter remind us of our humanity and our need for divine support as we navigate the complexities of faith and purpose.

2 Cor.12:7-8

"Therefore,[g] so that I would not become arrogant, a thorn in the flesh was given to me, a messenger of Satan to trouble[h] me—so that I would not become arrogant.[i] 8 I asked the Lord three times about this: it would depart from me."

Embarking on a prophetic journey can be both exhilarating and daunting, especially when trying to understand your unique role in the grand scheme. It's not uncommon to struggle with the intricacies of this path, as it requires a deep understanding of oneself and a connection to a higher calling. I empathize with this struggle, having dedicated over two decades to navigating this prophetic road. Much like the apostle Paul, I feel a solid duty to guide and support fellow prophets in their quest to unveil their true identities.

My journey has been akin to a lengthy molding process on the Potter's wheel, with each twist and turn leading me to the profound insights I possess today. However, I am acutely aware of the time and effort it took to reach this point, and I want to help others avoid the same arduous path. Therefore, I am driven to share my experiences and revelations with you, hoping they illuminate your journey.

My primary objective is to aid you in your spiritual growth, facilitate a deeper connection with God's grace, and impart the apostolic revelations I have received throughout my expedition. Like Paul, I consider myself a captive of Christ, bound by a divine purpose that transcends earthly limitations. The validity of my narrative is supported by legal documentation that remains accessible today.

Remember, you do not have to navigate this intricate path alone. Allow me to accompany you on this transformative journey, offering guidance and support. The book encapsulating my experiences is a testament to the lessons learned and the wisdom gained, serving as a beacon to illuminate your prophetic voyage.

The Test That Shifted Our Family's Compass

As a stay-at-home wife and mother, taking on the responsibility of homeschooling four children without state authorization has presented significant challenges in our daily lives. Despite these hurdles, we have cultivated a unique and independent educational environment for our children, allowing us to impart values and beliefs that resonate with our family's core principles. Navigating the complexities of homeschooling without official approval requires much creativity, dedication, and resourcefulness. We meticulously plan and execute each lesson to ensure our children receive a well-rounded education that encompasses both academic subjects and the moral and ethical teachings vital to us.

While driving one afternoon, we experienced a profound spiritual connection—a feeling like God was speaking to us. This awe-inspiring moment reinforced our belief in the path we have chosen for our family and served as a powerful reminder of the importance of faith and perseverance in the face of challenges. It reaffirmed our commitment to raising our children in a manner that honors our beliefs and values.

THROUGH THE TRIALS and triumphs of independently homeschooling our four children, we have discovered a deep sense of fulfillment and purpose in nurturing their minds and hearts. Each day presents an opportunity to educate them academically, shape their character, and instill the virtues that will guide them throughout their lives. While our unconventional approach to education has challenges, the rewards of witnessing our children grow and thrive in an environment rooted in love, faith, and family are immeasurable.

"WHAT'S KILLING you isn't what you see, but what you cannot see," God stated.

OUR FAMILY HAD JUST EXITED the military and settled in Broken Arrow, Oklahoma, believing everything would be great since God had led us there. As I reflected on what God was trying to communicate, I didn't fully grasp it then. I was unaware that God was upgrading me without my consent. I'm about to share a story based on an actual court case found at the Broken Arrow, Oklahoma, Courthouse. Although the names have been changed, the events are genuine. I struggled to figure out how to present this information without revealing the entire story, but I'm relieved there's no need to defame anyone involved.

MY HUSBAND and I served our country for over twenty years, but God called us to attend the School of the Prophets in Broken Arrow, Oklahoma, under Prophetess Vivian Sooner. We came eager to learn, although it was my second experience at a School of the Prophets; my first was in Frankfurt, Germany, under Bob Goodwin.

As we settled into this new area, we recognized many familiar faces within the faith community, and it felt as though we had arrived at the Church Utopia of the world. As a young couple who had chosen to leave the military and fully dedicate ourselves to serving God, we stepped into this new chapter with hope. Though we had never been to Oklahoma before and were unaware of church politics, our inexperience allowed us to trust and follow God's lead without hesitation.

After we arrived, we applied to a school, rented a beautiful home, and prepared to homeschool our four children. Eventually, we discovered that we were called to the School of Prophets led by Prophetess Vivian Sooner, who was associated with a prominent church in the area called Elevated Dimensions. Although we didn't know anyone there, we felt comfortable attending the church and began going regularly. Prophetess Vivian Sooner was the house prophet of the church, and her teachings resonated deeply with us.

After receiving a divine inspiration that stirred a deep calling, I felt compelled to dedicate myself to serving the Woman of God. Drawing from my experience as an armorbearer, I initially thought this new calling was just another routine task. Little did I know that it would lead me on a profound journey of faith and self-discovery.

My daily routine was meticulously organized to balance my responsibilities. Each morning, from eight until noon, I diligently homeschooled my children, instilling in them the values and knowledge I held dear. As they continued their studies after lunch, I embarked on my mission to support Prophetess Sooner. From 12:30 to 4:30 pm, I assisted her in various tasks, ranging from mundane errands to crucial appointments. Through this service, I sought to fulfill my duty and deepen my understanding of God's will and teachings.

My upbringing in a military family had ingrained in me the significance of discipline and unwavering focus on a mission. These qualities, honed through years of training and experience, became integral to my character and guided me in my service to the Woman of God. The lessons learned during this period became a lifelong habit shaping my actions and decisions.

Upon completing my tasks with Prophetess Sooner, I would return home to fulfill my role as a caregiver and provider for my family. However, unforeseen challenges awaited me two years later, as a tragic incident involving my mentor's husband thrust me into a tumultuous legal battle that shook the foundations of my faith. Amid this turmoil, the unwavering support and guidance of Arch Bishop Byron Ashen and my Spiritual Mother, Carolyn Cambridge, became beacons of hope and strength that sustained me through the arduous three-and-a-half-year ordeal.

Looking back on that challenging chapter of my life, I am grateful for the lessons learned, the resilience gained, and the unwavering support of those who stood by me in my darkest hours. The trials and tribulations I faced deepened my faith and strengthened my resolve to continue walking the path before me, guided by the light of God's grace and the wisdom of my mentors.

. . .

IT WAS a defining moment when I decided to break the tradition of keeping my dreams to myself and confide in the Prophetess Sooner. Despite my role as an armorbearer, which primarily involved assisting her with daily tasks and not discussing personal matters, I felt compelled to seek her insight on a dream weighing heavily on my mind. Unlike any other, this dream carried profound significance that I couldn't ignore.

AFTER YEARS of dedicated service and academic excellence, I had built trust with Prophetess Sooner, enabling me to step out of my comfort zone and share this intimate part of myself with her. The decision to open up about my dream was not made lightly; it was a testament to my deep respect and admiration for her wisdom and understanding of spiritual matters, especially dreams and their interpretations.

AS I RECOUNTED the details of my dream to Prophetess Sooner, I felt a sense of relief and vulnerability, knowing that I was entrusting her with something sacred to me. Although she could not provide an immediate interpretation, her willingness to listen attentively and offer her presence was a comforting reassurance. The timing of our conversation, just before her departure on vacation, seemed almost serendipitous, as if the universe had orchestrated our meeting for this precise moment. In that brief exchange, I shared a dream and a piece of my innermost thoughts and emotions, forging a deeper connection.

THE NEXT DAY, when Prophetess Sooner went on vacation, I received a call from Shay Biller, a local nail technician and a friend of Archbishop Byron Ashen. Shay offered me a discount on my nails, which I gladly accepted. Shay and her sister Monica, a psalmist, have always been my favorites. Monica is a fantastic and creative musician. My family loves it when the Pastor of Elevated Dimensions invites Prophetess Monica as a guest. As I sat down to get my nails done, Shay started a pleasant conversation but then abruptly asked...,

"Tell me about your dream involving Arch-Bishop Byron Ashen," she asked. "I have never dreamt about Bishop Byron," I replied. "Vivian mentioned you discussed a dream that included Arch-Bishop Byron Ashen."

I was overwhelmed by a surge of anger that consumed me like a raging wildfire. The events justified my long-standing avoidance of interactions with women, fueled by my perception of them as gossipy and chaotic. As a military woman, I have always prioritized substantive matters over trivialities, and this recent incident only reinforced my beliefs.

It soon became apparent that what I initially dismissed as harmless chatter was a carefully orchestrated scheme. I realized I needed to be cautious in my interactions, as the women involved, especially the instigator, Prophetess Sooner, appeared to thrive on creating chaos. To end her disruptive behavior, I confided in her about my dream, hoping it would calm the situation and repair the damage she had caused. Unfortunately, my efforts were in vain as I was subjected to a relentless barrage of questioning for an agonizing hour, all under the guise of verifying the authenticity of my words.

To my dismay, I was falsely accused of being betrayed by the woman I had been providing transportation for, despite never disclosing personal details about my family to Prophetess Sooner. It was clear that deceit and manipulation were at play, aimed at sowing confusion and tarnishing my reputation. I slowly recognized that Prophetess Sooner was orchestrating a campaign to damage my name among those she spoke with, creating an atmosphere of doubt and mistrust regarding my character. The treacherous web of lies and deceit she wove threatened to engulf me in a whirlwind of turmoil and uncertainty, leaving me to grapple with the repercussions of her malicious actions.

I FELT TRAPPED! I was getting my nails done when I suddenly had a strong urge to run away. However, being in my twenties, I didn't know how to confront authority figures, so I stayed and endured the situation, even though I should have left. Later that evening, I wrote a letter to Prophetess Sooner, expressing how hurt I felt when she lied about me to other leaders. Having learned communication skills during my time in the military, I could speak up for myself without being disrespectful or combative. I hoped my actions would lead to an apology and that things would return to normal. Prophetess Sooner was expected to return home the following day, so I waited for her response. She called me and invited me over to her place. When I arrived, she apologized, assuring me that there had been a misinterpretation. I believed her, given her divine position. She advised me to take the rest of the week off from my duties and let her handle the situation. We agreed to talk the following week.

I NEEDED to report to Prophetess Sooner for my assignment the following week. She was staying at the Holiday Inn off the highway, and I was instructed to meet her there. As I entered the building, George walked out of the elevator. I greeted him and was about to enter the elevator when he responded.

"YOU AND MY wife are cool, but I am not."

. . .

As he talked, I backed up, feeling a surge of unease creeping up my spine. Instinctively, I moved towards the courtesy phone to seek refuge in the familiarity of calling the Prophetess for guidance. However, George's sudden and aggressive actions caught me off guard before I could even press a button. In a swift motion, he grabbed the phone with my hand still grasping it, forcefully slamming it down on the table with a loud thud that reverberated through the room. The jarring impact sent a shockwave of confusion through me as he then turned towards me, his eyes ablaze with intensity. Without warning, he took the tips of his fingers to my shoulder, applying pressure that felt like a warning before jabbing me in the chest not once but three times in quick succession. The physical assault, combined with the unexpected turn of events, left me reeling, my mind struggling to process the sudden shift in dynamics. At that moment, a wave of disbelief and fear washed over me, rendering my thoughts momentarily paralyzed as I tried to make sense of the situation unfolding before me.

I began to scream, "Call the cops!" Call the cops!"

With a sinister demonic look, he said,

"The cops? Like you did last time, watch what happens," he exclaimed with a tone filled with a mix of fear and defiance.

His heart raced as he ran out of the hotel, the adrenaline coursing through his veins. The memory of the previous encounter with law enforcement is still fresh in his mind, a sense of urgency pushing him forward. Each step he took echoed his inner turmoil, the weight of his actions heavy on his shoulders. The city lights blurred as he sprinted through the streets, his mind a whirlwind of emotions and possibilities. Only time would tell what awaited him at the end of this chase.

. . .

THE YOUNG WHITE lady desk clerk, her panic mirroring my own, tried to calm me as she informed me that the police had been called. Just then, the elevator doors opened, and Prophetess Sooner, eager to understand the situation, rushed towards us. I recounted the incident to her, only to be met with a plea to keep the authorities out of it for the sake of her ministry. This unexpected request left me in internal conflict, torn between my sense of duty and the potential harm to her ministry. With a heavy heart, I began to distance myself from her, making my way to my car as she continued to implore me, her words echoing in my mind. "Please, don't call the police," she repeated, her voice tinged with desperation. "You're trying to hurt my ministry," she cried out, her distress palpable. I drove around aimlessly, grappling with the moral dilemma thrust upon me, trying to decide what steps to take next. The sense of unease and confusion lingered as I navigated the streets, searching for clarity amidst the chaos unfolding.

I KNOCKED on Carolyn Cambridge's door, as she lived just around the corner from me and was my husband's teacher at Prophetess Sooner's School of the Prophets. She answered the door, welcomed me in, and allowed me to sit and cry on her shoulder.

Then she asked me,

"What are you going to do?"

"I DON'T KNOW!" I replied.

"WHAT DO YOU MEAN, you don't know?" she stated.

"THEY ARE GOD'S LEADERS..." I struggled to speak.

. . .

"If you were still in the military and this happened, what would you do?" she asked.

"I'd press charges!" I responded.

"Stop looking at the God part, and do what you know needs to be done."

My phone rings, and it's Mother Dorian.

"What happened? Are you okay?" she asked.

"How did you hear about this?" I asked

"They're trying to control damage, so what will you do?" She asked.

"I don't know, I just need to think," I responded.

After I ended the call, I turned to Carolyn and informed her that I had to leave. She nodded, saying that I could always revisit her. My military training kicked in, and I began contemplating potential solutions to the problem. The big question was how to break the news to my spouse and what kind of reaction I should expect.

My mind was a whirlwind of conflicting thoughts and emotions as I exited the room. The weight of the situation bore down on me, making it hard to focus on a straightforward course of action. I couldn't shake the feeling of being torn between loyalty to my beliefs and the need to address the issue. Walking down the street, I replayed the conversation with Mother Dorian in my head. Her concern was palpable, and it only added to the pressure I was already under. The gravity of the decision ahead loomed large, and I knew that whatever choice I made would have far-reaching consequences.

ARRIVING HOME, I was greeted by the familiar sight of my spouse, unaware of the storm brewing inside me. How would they react to the news? Would they understand the situation's complexity, or would it drive a wedge between us? These questions swirled in my mind, intensifying the sense of unease that had settled in. As I sat down to gather my thoughts, I realized that this moment would define my actions and character. It was a test of courage and integrity, a challenge demanding a resolution aligned with my values. Taking a deep breath, I braced myself for the difficult conversations ahead, knowing that the path forward would require strength and determination.

AS I STEPPED into the familiar surroundings of my home, a heavy silence enveloped the air, rendering me speechless. Without uttering a single word, we found solace in each other's embrace throughout the evening, our unspoken thoughts echoing in the quietude. The next day, as the first light of dawn crept in, breaking the stillness, my phone pierced the tranquility with its insistent ring, revealing a call from Arch Bishop Byron Ashen. His urgent tone conveyed a sense of gravity as he instructed me to secure a restraining order following the filing of a police report, emphasizing the need for swift action before the clock struck one. The mention of Elevated Dimensions and the presence of Bishop Pearson awaiting my arrival filled me with hope and trepidation, as if divine intervention was unfolding before me.

. . .

GRATEFUL FOR THE GUIDANCE, I was determined to address the situation, yet the weight of the impending decisions loomed heavy on my mind. Despite the urgency of the Bishop's directive, I knew that transparency with my husband was paramount. Contemplating the best course of action, the early hour warranted my immediate attention to the legal proceedings. I intended to return home to share the events with my spouse before we departed for the church.

BY THE TIME the clock struck 10:30 that morning, I found myself back within the confines of my home, the gravity of the situation pressing upon me as I recounted the previous night's events to my husband. His stoic acceptance and unwavering support in the face of uncertainty spoke volumes, his silent acquiescence a testament to the bond that held us together. We prepared to embark on the journey with a sense of resolve, navigating the unknown with a shared understanding and unwavering trust in each other.

I ARRIVED AT ELEVATION DIMENSIONS, where Prophetess Sooner, Bishop G.L. Pearson, his assistant Pastor Cynthia Brooks, my husband, and I were in attendance. As we gathered around the table, Bishop Pearson avoided looking me in the eye and turned to my husband, asking him, "What do you think about all this?"

MY HUSBAND REPLIED, *"George called me last night and said he didn't want my wife serving his wife anymore, so I knew nothing else until today. I don't like it."*

I DROPPED my gaze to the floor, wondering why he had never said anything. Perhaps it was for the same reason I remained silent—I knew my husband was afraid of confrontation. Bishop Pearson responded, *"Well, it's in the courts now, so let them handle that matter."* Afterward, we all walked our separate ways.

Two weeks later, I had a dream in which God instructed me not to accept hush money. At that time, I needed to understand what that meant entirely. Soon after, Prophetess Sooner requested my presence at her office. When I arrived, she offered me six round-trip tickets to Hawaii with all expenses paid, stating that it was a reward for my service to her and her family. At that moment, I heard the Holy Spirit tell me, "Don't take the hush money." This was all the clarification I needed.

Our journey was not without its challenges. We lost some friendships we thought were strong, and others ended due to the pressure of choosing sides. However, in our time of need, God sent us unexpected allies. Strangers familiar with the couple's history appeared and offered their prayers, becoming our unexpected but much-needed support system.

Even in the most unexpected situations, transformation is possible. Bishop Pearson's assistant, Pastor Cynthia, initially believed that my family's presence at church was part of a plan to harm the city's famous Prophetess and her family. She preached this every Sunday to rally support. However, a conversation with her led to an apology and the beginning of a new, unexpected friendship. This experience showed us the transformative power of apologies and the potential for new beginnings.

One evening, while having dinner with my family, I received a call from Chanel, a student at the School of the Prophet. Chanel accused me of being responsible for Prophetess Sooner's removal from her position at the church. According to Chanel, a meeting had been called to discuss the removal of Prophetess Sooner from her role as House Prophet. Archbishop Byron Ashen had requested a follow-up conversation before he would agree to attend the annual conference.

I TOOK a moment to gather my thoughts and responded to Chanel, "ME, a no-name little person, told a BISHOP to REMOVE HIS chosen House Prophet from that position? Or would we—my family —retract our membership and $500 tithes elsewhere? You do know how silly that sounds, right?

THERE WAS a tense silence on the other end before Chanel finally replied, *"I am so sorry,"* and then hung up.

THE NEXT DAY, I called Pastor Cynthia and recounted the entire story to her. Silence hung in the air for at least a minute before she responded,

"NO ONE WAS in that office except for me, Bishop Pearson, and Karen Smalls. So, who called you and told you this?"

"*CHANEL*," I stated."

"SHE MENTIONED they were at a birthday party when the Prophetess addressed more than twenty people. I asked, 'Why would a Bishop do this?'

. . .

THIS INTRIGUING SCENARIO raises questions about the role of spiritual leaders in social gatherings and the impact of their words on the community. The birthday party, usually a casual and joyful occasion, contrasts sharply with the severe nature of the Prophetess's message, adding complexity to the situation. The presence of a Bishop further complicates the narrative, as it challenges conventional expectations of religious figures and their behavior outside of formal settings.

SHE STATED that Archbishop Byron Ashen would not agree to speak at the PSWAG Conference unless the issue had been resolved.

"THUS, the plan was to make you the scapegoat."

EXPLORING the motivations behind the Bishop's actions opens up a discussion about the intersection of personal beliefs, societal norms, and religious authority. It encourages reflection on the responsibilities and boundaries of spiritual leaders in various contexts, highlighting the complexities of human behavior and the dynamics within faith communities.

MY FAMILY and I decided to leave Elevated Dimensions to ensure our safety amid a legal case that dragged on for over three and a half years. Throughout this challenging period, we felt unfairly targeted by slander and propaganda, adding to the stress and uncertainty we were facing. Interestingly, about a year after we departed from the church, I received what I can only describe as a divine message from God, urging me to attend their annual Prophetic Conference. This revelation surprised me and left me feeling conflicted. However, reflecting on my time in Germany, where I learned to focus on the message of God rather than get caught up in unnecessary drama, I approached the conference with a similar mindset.

. . .

Despite my reservations, I decided to attend the event while keeping my reasons for being there to myself. It was an opportunity to reconnect with old friends unaware of my situation and observe how events would unfold within the church community. This experience taught me valuable lessons about resilience, faith, and the importance of staying true to my beliefs, even in challenging circumstances.

Every session, whether during the day or at night, was extraordinary. Each speaker had a message from God, but I preferred to sit at the back to avoid interacting with others. Although my out-of-state friends provided a refreshing perspective, the new students had already heard of me and knew what I looked like. When they spotted me, they reported my actions and appearance, making it clear that my presence made them uncomfortable. Nevertheless, I intended to stay focused, and the military training that had brought me to this point would endure.

The lunch layout was fantastic and beautiful, so I sat far to the right to avoid being seen. However, Prophetess Sooner noticed me, leaned over to the guest speaker, and pointed in my direction, revealing that I was the individual she was still in court with. Prophetess Fianna immediately scrutinized my existence and looked at me with disdain.

This event tested my emotional strength and ability to navigate uncomfortable situations. Despite the judgmental stares and whispers, I remained resolute in my decision to be present. The internal conflict between wanting to blend in and staying true to myself was palpable. Each moment felt like a battle between the urge to retreat and the determination to stand firm in my beliefs.

. . .

As the event progressed, I reflected on the significance of resilience and faith. My challenges only reinforced the importance of remaining grounded in my convictions, even when faced with adversity. The reactions of those around me reminded me of the power of unwavering faith in the face of skepticism and doubt.

Prophetess Fianna's piercing gaze bore into my soul as she scrutinized my every move, her expression laced with unmistakable disdain. Feeling a wave of discomfort wash over me, I politely excused myself from the company of my friends at the table, seeking solace in the sanctuary of the restroom. The cool tiles underfoot offered a momentary respite as I attempted to gather my thoughts and regain composure.

As I pushed open the door to the ladies' room, I saw a sight that made my heart skip a beat. Prophetess Fianna stood there, her imposing figure blocking my path. Her intense gaze locked onto mine, sending a shiver down my spine. I could feel the weight of her authority in the confined space as she uttered, "I k*now you're not leaving, are you? Because I have a word for you!*"

Caught off guard by her sudden appearance and the gravity of her words, I mustered a feeble response, "*No,*" my voice barely above a whisper, betraying a hint of apprehension.

The intense emotions and turmoil that I experienced at that moment were overwhelming. The writing on the wall seemed clear, indicating that I was left to fend for myself despite feeling like I was following divine guidance. As I gathered my thoughts and returned to my table, a sense of foreboding crept over me. I silently prayed for protection, seeking solace in my faith as tensions rose in the room.

. . .

THE SITUATION ESCALATED QUICKLY as the guest Prophetess unleashed a barrage of curses upon me, attacking not just me but also my family, my marriage, and even my call as a prophet. Despite the onslaught, I stood my ground, unwavering in my resolve. Like the Prophet Jeremiah, I maintained a steadfast demeanor, my expression resolute despite adversity.

AS THE SERVICE DREW CLOSE, a numbness settled in, enveloping me in disbelief. Concerned friends approached me, offering support and a listening ear, but I chose to mask my inner turmoil with a smile, bidding them farewell as they departed for Kansas. Observers scrutinized my every move, attempting to decipher the emotions hidden beneath the facade I presented.

WITH A HEAVY HEART and tear-filled eyes, I turned to God, pouring out my anguish and frustration. The weight of the situation bore down on me, pushing me to a breaking point. In that moment of vulnerability, I solemnly vowed to confront Prophetess Fianna, fueled by anger and determination.

AS I MADE my way down the seemingly endless hallway, a whisper from the Holy Ghost resonated within me, offering a glimmer of hope amidst the darkness. The assurance of ten thousand angels preparing the way for me provided comfort and reassurance, instilling a newfound strength within my weary soul.

WITH A RENEWED SENSE OF PURPOSE, I steeled myself for the confrontation ahead, ready to face the challenges that lay in wait. The resolve to confront Prophetess Fianna loomed large in my mind, a testament to my unwavering faith and determination to seek justice in the face of adversity.

Soul Crushing Reality

Driving home that day, I felt betrayed, questioning why God allowed such events to unfold despite guiding me. Questions flooded my mind, from feeling set up to being mistreated by those I served. The turmoil of being accused and standing alone against scapegoating weighed heavily. The journey of integrity training seemed futile as I faced unjust treatment. The complexity of relationships and actions left me wondering how to move forward with respect and dignity.

CAROLYN, a constant presence in my life from the beginning, graciously opened her home to me. Overwhelmed by my emotions, I sought comfort in her company, eager to share the burdens weighing on my heart. As I expressed my feelings, a question lingered: *"What had I done to deserve this?"*

HER RESPONSE SURPRISED ME. She paused for a moment of contemplation before asking, *"Will you forgive me*?"

Confused, I asked, "*Forgive you for what*?"

SHE HESITATED BEFORE REVEALING, "*I held negative feelings towards you. I allowed myself to be influenced by the negativity surrounding you, which all stemmed from the woman you faithfully served in recent years."*

THIS REVELATION STRUCK a chord with me, unraveling a complex web of misunderstandings and misplaced judgments. It was a pivotal moment that highlighted the power of perception and the impact of rumors on relationships. In that instant, a journey of forgiveness and understanding began, reshaping the dynamics of our bond and emphasizing the importance of looking beyond hearsay to uncover the truth within each other.

FOR A MOMENT, I felt disheartened. The truth was that the Prophetess, who was supposed to guide and teach me, was inadvertently causing harm to the nail technicians and her students, teachers, staff, and paid speakers. I sat with the harsh reality of the situation, feeling sick, hurt, and shocked all at once. Our family had left a controlled military environment and entered a chaotic religious world. This experience was the final straw for me. At the time, Carolyn was just a teacher at the School of Prophets, but after this, we became friends. Her profound impact on my life earned her the titles of Confidant, Pastor, Prayer Partner, and Spiritual Mother.

AS I CRIED out to Carolyn, she said, "*You don't deserve this, and we all know how her husband is and what he's done to other women. But you were the first to stand up to the abuse! Everyone knows you're not lying, but they're too afraid to speak. I will not be silent anymore. Please forgive me, and I am sorry.*"

"YOU'RE FORGIVEN," I replied.

AS I BEGAN WRITING this book, God directed me to Paul's opening scripture to reveal an important truth: where there is a revelation, there will also be a thorn. For prophets, this thorn often manifests as rejection. However, I initially failed to grasp that prophets are also called to confront and contend with the abuse of authority and manipulation.

. . .

God emphasized this message to me: "*This whole ordeal is called Germ Warfare, and you must learn it quickly.*"

This revelation shook me to my core. It was a call to arms against the misuse of power and manipulation that had permeated the religious circles I found myself in. The weight of this responsibility settled heavily on my shoulders, knowing that I had to stand up against the darkness that had infiltrated even the most sacred spaces.

Carolyn's words echoed in my mind, reminding me that silence was no longer an option. I had to be a beacon of truth and justice, even if it meant facing backlash and resistance from those who preferred to look the other way.

With each word I penned in my book, I felt a sense of liberation and empowerment, knowing I was bringing light to the shadows that had clouded the spiritual realm. The journey ahead was daunting, but I was fueled by a sense of purpose and a deep-seated conviction that I was meant to be a warrior against injustice.

As I delved deeper into Paul's teachings and Carolyn's lessons, I realized that my path was not just about personal growth but about standing up for what was right, even in the face of adversity. The thorn of rejection was a small price to pay for the opportunity to create change and expose the toxic patterns that had long been overlooked.

. . .

God's message resonated within me, urging me to embrace the role of a spiritual warrior in the battle against abuse and manipulation. The journey ahead would be challenging, but I was prepared to walk this path with courage and unwavering faith, knowing I was not alone in my quest for truth and justice.

2

GERM WARFARE BASICS

Although we still have some way to go, we will continue uncovering the story's essence. Let's delve deeper into the intricate world of germ warfare. Webster's Dictionary defines this form of warfare as deliberately introducing disease-causing microorganisms into enemy territory during conflict. The very nature of germ warfare presents a unique challenge due to its invisibility, making it a silent yet potent weapon.

Exploring the significance of germ warfare reveals that this method poses a grave threat not only to military forces but also to civilian populations. The insidious nature of biological agents used in germ warfare can spread rapidly and indiscriminately, causing widespread devastation and chaos. The use of such tactics raises ethical concerns and questions regarding the boundaries of warfare and the protection of human life.

Key takeaways from this definition include the realization of the clandestine and insidious nature of germ warfare. It underscores the importance of vigilance and preparedness in detecting and countering such threats. Understanding the implications of germ warfare not only highlights the need for international cooperation but also emphasizes that stringent measures are crucial to prevent the misuse of biological weapons. It's a global issue that requires a united front.

As we continue exploring the realm of germ warfare, we must remain vigilant and proactive in addressing this evolving threat. It's not a passive task but an active one that demands our attention. By laying the groundwork and gaining a deeper understanding of the significance of germ warfare, we can better appreciate the complexities and implications of this form of warfare in our modern world.

Key Highlights

- *Germ Warfare is the Deliberate Attempt to Destroy*
  - *Objective: To kill.*
  - *A Contaminating Mixture Is Death*
  - *An Attack on an Enemy or Opposing Force*
  - *Seen in War Zones*
  - *Weapon Of Surprise*
  - *Strategically Planned*
  - *Clever Trap*
  - *Promotes Disease*

During the Vietnam War, the use of Agent Orange by both the Vietnamese and American forces had long-lasting consequences that extended far beyond the battlefield. The decision to spray this toxic herbicide on marijuana fields not only aimed to disrupt the enemy's food supply but also inadvertently exposed American soldiers to its harmful effects. Agent Orange contains dioxin, a highly toxic compound that has been linked to various health issues, including cancer, congenital disabilities, and neurological disorders. For the soldiers who were exposed to Agent Orange during the Vietnam War, the repercussions were severe and enduring. Many veterans developed debilitating health conditions, such as skin rashes, respiratory problems, and gastrointestinal issues, which persisted long after their return home. The toxic legacy of Agent Orange also affected the offspring of these veterans, leading to a new generation of individuals grappling with genetic mutations and chronic illnesses. The use of Agent Orange in Vietnam not only ravaged the landscape and ecosystem but also left a dark legacy of suffering and uncertainty for those who were exposed to its toxic effects. The revelation that a chemical weapon designed to destroy vegetation was unknowingly used on marijuana fields underscores the complex and tragic nature of modern warfare, where the consequences of such actions reverberate for decades to come.

- Lost or hindered visibility
    - Loss of alertness
    - Loss of mental state
    - Loss of physical strength

War is inevitable!

As human beings, we are all engaged in a continuous internal battle within our consciousness's depths. This conflict may not always be visible on the surface, but it nonetheless influences our thoughts, actions, and perceptions. Regardless of our awareness or chosen stance, we are all warriors in this complex battlefield of existence.

Our individual beliefs and ideologies may seem trivial in the grand scheme of this internal struggle. Yet, we must equip ourselves with the necessary tools and insights to navigate this turbulent terrain.

Understanding the fundamental truths about this internal conflict is crucial for our growth and evolution as individuals.

In all its forms, war has a set of undeniable realities that we must confront and accept. Through this confrontation, we can unravel the complexities of our inner struggles and pave a path toward self-discovery and enlightenment. Embracing the challenges of this internal war can lead us to deep insights and transformative experiences that shape our essence.

- Casualties will accrue
- Good people die, as do the bad.
- War doesn't care about your color, only cause and outcome.
- Equipped to fight or not, you'll become either a warrior or a casualty.
- War seeks to win at all costs, including the cost of life.

I want to share a military perspective that sheds light on our daily enemy. "war" signifies a conflict or division between countries or sides. At the same time, "zone" refers to a specific location or district. Combining these two words shows us that a war zone can occur anywhere. Unfortunately, the term "war zone" has become all too common in our lives.

Germ Warfare is the intentional attack and contamination of an unseen enemy to destroy someone or something. This strategy is designed to be (1) invisible, (2) intentional, and (3) toxic enough to cause death. In the past, the term has been used to describe combat or war zones. However, in today's world, we can consider the world as a whole to be a war zone, whether it's the suburbs, low-income areas, or even the wealthy neighborhoods. War is inevitable!

AN EXAMPLE of a war zone is the conflict between the United States and Iraq, which has become well-known. Iraq is the immediate war zone, but the US has also become a war zone within itself. This happened when terrorists invaded the country's borders to attack it. We can see that the time between Vietnam, WWI, WWII, and the first and second Iraqi invasions was only a moment in history. In one moment, peace ended, and a war zone was created. However, if we look at our history, we discover that war zones have existed within our communities for years. These war zones were created from our past judgments towards races, cultures, beliefs, hate crimes, and political agendas. Whether we agree or disagree, conflicts have always existed. For a war zone to exist, there must first be a conflict that leads to a war. In ancient times, there were two types of wars:

ON ONE HAND, wars were fought for territorial expansion, resources, power, or dominance. These wars were often driven by the desire for control and supremacy over land or people. On the other hand, wars were fought for ideological or religious reasons. These conflicts stemmed from differing beliefs, values, or interpretations of faith, leading to violent confrontations and battles.

WAR ZONES HAVE EVOLVED and adapted to society's changing landscapes throughout history. From traditional battlefields to modern urban environments, the concept of a war zone has expanded to encompass various forms of conflict and violence. In today's interconnected world, the boundaries of war zones have blurred, with threats emanating from both physical and digital realms.

. . .

As we navigate the complexities of a world where war zones can manifest unexpectedly, it is crucial to remain vigilant and proactive in addressing the root causes of conflicts. By promoting understanding, dialogue, and cooperation, we can strive towards a future where the term "war zone" is relegated to history books and peace prevails.

(1) Wars of expense and expansion, and

(2) Holy or Religious wars, which were unlimited, compulsory, and affected entire nations.

Many people believe that the war in Kuwait began in 1991 when US troops first entered the country. History books state that twelve years later, in 2003, the US entered Iraq and captured the notorious dictator Saddam Hussein. However, it is inaccurate to say that there was no war zone in Iraq during the twelve years between these two events. Tension and violence remained throughout the region, best described as a ceasefire. Efforts were made to find a more effective resolution to avoid a massive military operation, but unfortunately, the conflict resumed twelve years later. Therefore, the idea that no war zone existed for those twelve years is incorrect.

The suffering of the Iraqi people began long before the US intervention. The US became involved after Iraq attempted to invade Kuwait, a situation of interest to both the US and the international community. While the US was primarily focused on securing oil, it is essential to acknowledge that the same resources were also vital to other countries. Kuwait had the right to seek assistance from the United Nations, and the US was one of the nations that responded.

Often, we hear about the terrible events that occurred in Iraq, the devastation caused by war, and the loss of life resulting from US actions. However, we frequently overlook the various violent acts that prompted our intervention in the first place.

BEFORE OUR INVOLVEMENT WITH IRAQ, Saddam Hussein, the dictator, was responsible for the deaths and torture of more than a million women and children. He executed thousands and detained many others based solely on suspicions without any evidence or charges. People were pronounced guilty without any fair trial, taken from their homes, and sent to prison, never to be seen again. Reports also detail how Saddam's sons would intrude on weddings, assault and rape the brides in front of their husbands, and then kill them if they spoke up. In addition to all this, Saddam had ambitions to take over Kuwait. When faced with a stronger adversary, the logical choice is to seek assistance from a more powerful ally to defend oneself. This is precisely the purpose of the United Nations!

FOLLOWING the events of 9/11, political considerations were set aside in favor of seeking retribution. The nation united to demand justice after realizing that an enemy had attacked us on our soil, violating us in front of the world. The shock was profound! Despite our differences in tactics, religious beliefs, and values, the younger generation of our adversaries seemed more adept at handling warfare and adversity than many adults in our military. Why is that?

FROM THEIR EARLY INFANCY, fathers took their youth and began to teach them to defend by any means necessary. Yes, the methods are extreme, and yes, we wouldn't dare think of doing the same thing to our youth, but maybe we should. These children are prepared to die for their family, religion, and country at a moment's notice while we divide our nation and kill each other. Each day we live, opposing forces surround us, using various strategies to ensure that our destinies don't come to pass. Germ Warfare is so decisive because it seeks to kill our identity.

Over the following chapters, let's focus on several topics of interest:

- The Christian Battlefield
- Why Identity Matters
- Components Of War

3

CHRISTIAN BATTLE GROUND

Within the context of the Church—traditionally associated with peace and compassion—the notion of war can be particularly unsettling. How does one reconcile the teachings of love and forgiveness with the brutality of armed conflict? This internal struggle highlights war's complexity, transcending physical battles, and encompasses moral, ethical, and spiritual dimensions that challenge our understanding of humanity.

Nevertheless, it is imperative for us, as human beings, to confront the stark realities of our world, including the presence of war and conflict. By acknowledging the existence of such challenges, we not only gain a deeper understanding of the complexities of human nature and the intricacies of our shared existence but also open ourselves up to the transformative power of these acknowledgments. Through introspection and reflection, we can glean valuable lessons about resilience, empathy, and the enduring power of hope in the face of adversity. In delving deeper into the truths surrounding war and its impact on individuals and societies, we can cultivate a greater empathy and understanding towards our fellow human beings. The

Church plays a significant role in this process with its teachings of love and compassion. Engaging with these complex and uncomfortable truths can foster a more compassionate and inclusive worldview that transcends the boundaries of religion, culture, and ideology. Through this process of introspection and dialogue, guided by the principles of the Church, we can begin to unravel the complexities of war and strive toward a more peaceful and harmonious coexistence.

The Real Battlefield

Read Gen.2:1-2,

*"Thus the heavens and the earth were **finished (3615)**, and all the **host (6635)** of them. "And on the seventh day God **ended (3615)** his **work (4399)** which he had **made (6213)**; and he rested on the seventh day from all his work which he had made."*

Let's Word Study This Scripture;

• **Finished (3615)**--*kalah*: to be complete, at an end, finished, accomplished, or spent

• **Host (6635)**--*tsaba*: army, war, warfare

• **Ended (3615)**-- *kalah*: to be complete, at an end, finished, accomplished, or spent

• **Work (4399)**--*melakah*: occupation, work

• **Made (6213)**--*asah*: accomplish

*"Thus, the heavens and the earth were **finished (3615)**, (to be complete, at an end, finished, accomplished, or spent) and all the **host (6635)** (army,*

war, warfare of them. "And on the seventh day, God ***ended (3615)*** *(to be complete, at an end, finished, accomplished, or spent his* ***work (4399)*** *(occupation, work) which he had* ***made (6213);****(accomplish) and he rested on the seventh day from all his work which he had made." (Gen.2:1-2)*

THE BATTLEFIELD for every believer starts here! The word "finished" signifies an expected conclusion to everything that will happen; God has already foreseen the outcome. The next word to focus on is "host" (H6635). Interestingly, we can gain deeper insights if we examine this word more closely using Strong's concordance.

WAR (41X), army (29x), battle (5x), service (5x), appointed time (3x), warfare (2x), soldiers (1x), company (1x), miscellaneous (5x).

THE TEXT DESCRIBES a battlefield or warzone during the seventh day of rest, symbolizing warfare's conclusion or prophetic end. It suggests that since the fall of the angels and leading up to Ephesians 6:14, there has been an ongoing battle against an unyielding enemy. Soldiers are depicted as being engaged in this conflict, confronting various hardships.

IN NUM 4:23;

"FROM THIRTY YEARS old and upward until fifty years old shalt thou number them; all that enter in to perform the ***"service", H6635*** *to do the work in the tabernacle of the congregation."*

THE TERM "*SERVICE*" in the above text denotes "host," as mentioned in Gen. 2:1-2. However, in Numbers, "host" refers to a duty and implies a military obligation. This duty is not just a simple task but a commitment that requires dedication and sacrifice. When we delve deeper into the service concept, we realize that it encompasses various responsibilities and roles individuals undertake for multiple purposes. People who have served on active duty understand that specific assignments are known as *"hardship tours."* This term carries a weight of challenges and difficulties that individuals must endure during service. For instance, a soldier stationed in South Korea would be on a hardship tour. Such soldiers are posted with the understanding that their dependents (family members) cannot accompany them on the tour. This separation adds another layer of complexity to their service, highlighting their sacrifices for their duty.

THE POINT IS that being stationed in South Korea requires more attention from all those assigned to the region. It demands heightened awareness, adaptability, and resilience from the individuals serving in that specific location. The region's environment, culture, and circumstances shape the nature of their service, making it a unique experience that tests their capabilities and character.

ADDITIONALLY, it should be noted that hardship tours of this nature have a fixed time frame or duration. This temporal aspect adds a sense of urgency and impermanence to the service, emphasizing the transient nature of their assignment. The word of God tells us that our existence on earth is similar to a hardship tour and has a specific period. Just like soldiers on a mission, we are tasked with fulfilling our duties and responsibilities within the time allotted.

GENESIS 2:1-2 offers an intriguing perspective on the term "host," which enhances our understanding of how God is portrayed as associated with war, soldiers, and armies. The terms "angels," "host," and "works" relate to angelic activity. In Hebrew, the word "angel" means "messenger" (malak), while "host" translates to "tsaba" (sa-ba) and denotes a large group of entities organized for battle, essentially an army. The term "works" in Hebrew is "maaseh," which signifies an action or a product. Angels function as messengers, sent with instructions, information, and commands, reflecting their role in angelic activity. Since the beginning of time, angels have fulfilled God's will, following His directive and establishing His dominion (Zech. 5:1-4);

Then I turned, lifted my eyes, and saw a flying roll.

2 And he said unto me, What seest thou? And I answered I see a flying roll; the length thereof is twenty cubits, and the breadth thereof ten cubits.

3 Then said he unto me, This is the curse that goeth forth over the face of the whole earth: for every one that stealeth shall be cut off as on this side according to it; and everyone that sweareth shall be cut off as on that side according to it.

4 I will bring it forth, saith the Lord of hosts, and it shall enter into the house of the thief, and into the home of him that sweareth falsely by my name: and it shall remain amid his house, and shall consume it with the timber thereof and the stones thereof.

The concept of warfare transcends the simplistic notions of good and evil, as all entities involved are referred to as "hosts," existing since the beginning of time. The term "host" implies a defined duration, which ultimately comes to an end (Genesis 2:1-2). The poetry of God is undeniable; it foretells that not only did He create the world, but He also anticipated that His people would endure hardships on earth. However, in the end, they will triumph.

Therefore, the reward is not given to those who are merely well-prepared or the most eloquent preachers but rather to those who persevere. We are active-duty soldiers assigned to earthly missions for a specific period. Humans are called to serve this world under unpredictable circumstances and for a period that only God knows.

As active-duty soldiers, we learned we had a specific job. We cherished our family time when the day ended and even went on vacations together. However, the difference with being a soldier is that we can be called to duty without prior notice. We can't call in sick or give any excuses. We are soldiers first and foremost, and we signed up for a mission that demands our presence, no matter what. God was using the military to teach me the importance of having a mission and understanding my calling. It's crucial to comprehend one's mission because failing to do so can lead to premature death on the battlefield. In the Vietnam War, more young people died than old ones. They were inexperienced souls, drafted into service without guidance.

I realized that God had a plan for my life that went beyond my understanding. To fulfill my purpose, I needed to acknowledge that Earth was not my permanent home and that I had been placed here to complete a specific assignment. Despite the challenges and difficulties that would come my way, I was determined to persevere. I stopped relying solely on books to guide me and instead began living out the truth of what it meant to fulfill my purpose. As I gained a deeper understanding of the reality of the battle I was in, I realized the importance of understanding my identity and how that would help me succeed in my mission.

Tulsa was the place where I began my Prophetic Office Training. It was the second school of Prophetic Training that I attended, after the one in Frankfurt, Germany, led by Bob Godwin. Through my time in Tulsa, I gained a practical understanding of the Office of the Prophet, which went beyond mere theory. Before we moved to Tulsa, my family and I had been active members of various ministries, both overseas and in the US, serving as diligent servants, elders, and armorbearers. We were always loyal and faithful, paying our tithes and affiliating with church staff. However, the move to Tulsa marked a new level of spiritual warfare for us.

TRANSITIONING to Tulsa brought with it a profound shift in my spiritual journey. The Prophetic Office Training I underwent in Tulsa was a continuation of my previous education in Germany and a transformative experience that deepened my connection with my faith and purpose. The teachings and practical insights I received in Tulsa opened my eyes to a new dimension of spiritual understanding and warfare.

AS I DELVED into the training in Tulsa, I challenged myself in ways I had never experienced before. The spiritual battles intensified, requiring me to rely even more heavily on my faith and understanding of my identity as a servant of God. Each lesson, each encounter, and each moment of reflection in Tulsa shaped me into a more robust, more resilient individual, ready to face the trials and tribulations that came with fulfilling my divine assignment.

MOREOVER, the move to Tulsa marked a geographical shift and a significant turning point in my spiritual growth. The community I found in Tulsa, the mentors I encountered, and the challenges I faced all played a crucial role in shaping me into the purpose-driven individual I am today. The spiritual warfare I encountered in Tulsa was not a hindrance but a stepping stone towards a deeper understanding of my calling and a more substantial commitment to fulfilling God's plan for my life.

. . .

For example, immediately after the assault, Archbishop Byron Ashen instructed me to call the police, file a report, and obtain a restraining order, which I did. This ordeal was unfamiliar to me; I was trusting individuals who didn't know me over those who were supposed to support my development but instead sought to harm me. No one seemed to notice how uncomfortable this situation was; all people cared about were the lies presented as truths.

The day came when I had to return to court to determine if the temporary restraining order would be made permanent. Prophetess Sooner discreetly passed a piece of paper to their church lawyer, who approached me on the stand and began to ask questions like:

"Isn't it true that you came to Tulsa to destroy these good people?"

"Doesn't this application indicate that our last church didn't even approve of your coming here?"

Wait! What? These people had taken the School of the Prophet's application form and used it to determine that I had planned on some mission to destroy their School (when I didn't know who they were). Our previous Pastor, whom we had served for 7 years, didn't even want our family to leave the ministry because of our servitude, so they disapproved of us going, which is why we never had a letter of recommendation. Of course, all this was known before being accepted. However, this was the extent to which they were willing to go to save the sin and actions of an abuser. Today, many who hide behind the name of Pastors are abusing God's sleep, and the members, staff, and others are quick to rally support, building illusional lies to protect their sin over the truth of their actions.

. . .

After the traumatic assault, the feeling of being misunderstood and unsupported intensified as the focus shifted from the truth to fabricated lies. Prophetess Sooner and the church lawyer's involvement added layers of complexity. The lawyer's accusatory questions, insinuating destructive intentions and misrepresenting past circumstances, painted a distorted picture of reality.

The misuse of the School of the Prophet's application form to construct false narratives about the individual's motives revealed a disturbing willingness to manipulate facts to protect the abuser. The lack of understanding and empathy from those in positions of authority further highlighted the toxic dynamics at play.

This ordeal shed light on a concerning trend: Some individuals misuse their positions of power within religious institutions to perpetuate abuse and shield wrongdoers. The narrative spun to justify unjust actions, and the willingness of others to disregard the truth showcased a systemic issue that needs to be addressed.

We were naive, and while we recognized that the church wasn't perfect, we saw it as a group of organized individuals who misused God's name to support their self-serving ambitions. These were clear examples of witchcraft in human form! They perceived their influence dwindling and decided to eliminate what they assumed was the enemy. But how can we harm those we have abused? How do you mistreat people and then place the blame on them as if they somehow asked for it? This misdirection helps me today as I interact with those who have experienced church hurt. When I walked out of that courtroom, I felt as if all my faith in church leadership had been stripped away after witnessing what I did. This false prophetess was openly practicing witchcraft and poisoning other prophets, and all of them fell victim to her lies.

. . .

As I sat in the parking lot that day, God reminded me yet again of what He spoke in Germany;

"Rejection will be your portion. A prophet is never accepted, but obedience is required no matter what and despite the rejection."

The Holy Spirit reminded me that He is my guide and counselor compass, so Germ Warfare was right up my family's military alley. It was time to see the truth of the battlefield from the inside out. This information will help you fight, strategize, and counteract the enemy's plans.

The Holy Spirit has always served as my source of guidance and wisdom and recently brought to my attention a profound realization: that the battle I am engaged in is not one of personal choice but rather a fundamental aspect of my very being. With the Holy Spirit as my constant companion, providing me with counsel and direction, engaging in Germ Warfare - a challenge that aligns perfectly with my family's long history of military service - took on a new significance. This revelation prompted me to delve deep into the battlefield's intricacies, gaining insights crucial for developing effective strategies and thwarting the adversary's schemes. Armed with this newfound understanding, I am better equipped to face the trials ahead with courage and resilience, knowing that the Holy Spirit will continue illuminating my path and empower me to emerge victorious in spiritual warfare.

4

WHY IDENTITY MATTERS

After finishing high school, my future felt uncertain and overwhelming. As summer approached, I was bombarded with thoughts about what came next, but nothing seemed to spark my interest. While visiting my friend Richard, he asked if I would like to join a young man named Charles—nicknamed Chucky—since we were the same age. Richard, a veteran and a local boxer, was helping Chucky prepare for the Armed Services entry test. Chucky had failed the test before, and Richard thought having me join would help him feel less intimidated, so I agreed.

WE ARRIVED at the Military Entrance Processing Station (MEPS) in Philadelphia and were directed to the testing room, where more than fifty people were waiting their turn. The atmosphere was tense, filled with a mix of nervousness and determination. The room buzzed with anticipation as each individual awaited their chance to prove themselves. The test itself was rigorous and challenging both mentally and physically. As the hours passed, the moment's weight settled in, each minute feeling like an eternity.

. . .

Chucky's expression grew somber as the results were announced; disappointment and frustration were evident. In contrast, I felt a surge of relief and accomplishment when I learned that I had passed. The recruiters wasted no time approaching me, presenting many opportunities that suddenly lay before me. Richard, a pillar of support throughout the process, had to leave with Chucky, leaving me to navigate this new chapter independently.

By Richard's return, my fate had been sealed—I had signed up and committed to joining the United States Army. The reality of my decision hit me as I realized that within thirty days, I would embark on a journey that would transform me in ways I could never have imagined. When the day arrived, as an 18-year-old girl from Philadelphia, I found myself back at MEPS, ready to be shipped off to Fort McClellan, Alabama. The mixed emotions were palpable—fear, uncertainty, and excitement for the adventures ahead.

Stepping into the unknown, it felt like the world was at my feet, brimming with possibilities and opportunities. The nine following weeks were a whirlwind of challenges, growth, and transformation. As I transitioned from a civilian to a soldier, the experience was profound, shaping me into a person I never knew I could become. I brought new lessons, experiences, and a newfound sense of purpose each day. As I navigated the training, camaraderie, and discipline of military life, I discovered strengths within myself that I never realized existed.

To better understand military terminology, it's essential to explore the specific meanings of certain words and their implications. A civilian is defined as a person who is not actively involved in the armed services or law enforcement. However, the shift from civilian to soldier is a profound and transformative experience for many individuals. When one takes the solemn oath of allegiance, a fundamental change occurs—marking a symbolic departure from civilian life and an entry into military service. This transition initiates a rigorous journey toward becoming a competent soldier.

Embarking on this path involves intense training and discipline, which shape individuals into capable armed forces members. The training period is crucial, as recruits acquire essential skills, knowledge, and values vital to their roles as soldiers. Even after initial training, a soldier's journey is continuous growth and development. Each day brings new challenges and opportunities for learning, refining skills, and building camaraderie within the military community.

Upon finishing their training, soldiers experience a sense of pride and accomplishment, knowing they have embraced a noble profession dedicated to serving and protecting their country. Although the training phase represents a significant milestone, it also marks the beginning of a lifelong commitment to uphold the values and responsibilities of a soldier. Therefore, the transition from civilian to soldier is not just a change in status but a profound transformation that shapes one's identity and purpose within the military hierarchy.

In civilian life, identity is often viewed as a blend of affiliations derived from various groups, such as family, ethnicity, and occupation. However, identity may be perceived differently in the military, especially for long-term soldiers dedicated to their careers. Some argue that military identity constitutes a social identity that emerges as civilian identity becomes less prominent due to military enculturation, eventually becoming an integral part of an individual's self-concept.[1]

. . .

At birth, identity is shaped by inherited traits passed down from previous generations. As individuals grow and evolve, their identity is influenced by genetic makeup and environmental factors. The Chronosystem refers to the comprehensive set of environmental changes that impact a person's development throughout their lifetime. This includes significant life transitions, such as starting school or getting married, and non-normative life events, like divorce or relocating to a new home.

Scientific Perspective

Urie Bronfenbrenner, the Creator of the Ecological Systems Theory, proposed that an individual's development is influenced by a range of interconnected environmental systems that vary from immediate surroundings to broader societal structures. Bronfenbrenner categorizes these systems into five types, defined as follows:

Microsystem: The first and most crucial level of Bronfenbrenner's theory, the microsystem consists of the direct contacts a child has in their immediate environment, including parents, siblings, teachers, and school peers.

Mesosystem: This level encompasses the interactions between the child's microsystems, such as the relationships between the child's parents and teachers or school peers and siblings.

Exosystem: This incorporates other formal and informal social structures that influence the child, even if the child is not directly involved. Examples include neighborhoods, parents' workplaces, parents' friends, and mass media.

. . .

Macrosystem: This component focuses on how cultural factors impact a child's development. Socioeconomic status, wealth, poverty, and ethnicity are included. The macrosystem differs from the microsystem in considering broader societal influences rather than specific environments.

Chronosystem: This system consists of all the environmental changes that influence development over a lifetime, including significant life transitions and historical events. Examples include starting school, marriage, and non-normative life transitions such as parental divorce or moving to a new home.

Our belief systems, emotional choices, and priorities shape how we approach life. These aspects may seem like defining identities, but they are defining moments that contribute to our life experiences. When faced with unexpected trauma, we might be tempted to view it optimistically, believing that it will enhance our identity. However, keeping our achievements separate from our sense of self is essential. Our identity is like an onion that must be peeled back layer by layer, revealing our challenges, accomplishments, disappointments, and defeats. We can only discover what truly matters at our core by closely examining all these layers. Even then, we must remain open to ongoing growth and evolution.

The mind of the Creator is beyond our comprehension. Until we return to Him, we will remain partially undefined and continually explore who we are called to become. God uses the world's foolish things to confound the wise, and this discovery often remains hidden from the ordinary person but is revealed as needed. Books on self-worth and identity are in high demand today as many search for answers. Even if we lived to be sixty or more, we would still be discovering ourselves! As a soldier, one quickly learns that we constantly gather knowledge, information, and experiences. This growth can compensate for areas where our development may have been hindered. While I enjoyed my time, I always strived to be the best and looked forward to surpassing the limitations of the system that raised me.

Identity

Unfortunately, many people made assumptions about me and my family without understanding our background and values. As a military family, we have a strong sense of honor, loyalty, and commitment, reflected in our integrity. I must stay true to myself, as my beliefs are essential to who I am.

Some individuals have questioned how I managed to endure being attacked in Broken Arrow, Oklahoma, at one of the largest churches in the area. Many assumed that our family's presence at Higher Dimensions was meant to cause harm to the church leadership. Others accused us of being mean-spirited for not leaving the church but instead choosing to stay and continue our support through tithes.

Many fail to understand that we are a military family. We are deeply rooted in honor, loyalty, and commitment, so our display of integrity is not an act; it is an essential part of our identity in Christ. God empowers us to fulfill His work through the guidance of the Holy Spirit for His glory. While we all inherit a spiritual identity, it takes time to mature mentally and fully embrace it.

I started studying German warfare, but God reminded me to see the battlefield with spiritual eyes instead of just physical ones. He also reminded me that my first allegiance should always be to Him, just like in the military. Soldiers understand the importance of fulfilling their mission, and as a veteran, I know that I must complete the assignment given to me, no matter what. Unfortunately, when veterans like myself remain focused on the mission in the local church, we are sometimes accused of being cold or untouchable. However, this couldn't be further from the truth. We are loyal and dedicated, and we take our assignments seriously. Ironically, the church claims to be in the Army of the Lord, yet many lack understanding of military matters such as friendly fire.

Being part of a military family instills a sense of discipline and responsibility that transcends mere duty. It shapes our character and influences our interactions with others. The values of honor, loyalty, and commitment are not just words to us; they are the pillars of our identity and the guiding principles of our actions. Embracing these values is a continuous journey of growth and self-discovery as we strive to align our spiritual identity with our everyday lives.

As we navigate life's challenges, we are reminded of the need to view the world through spiritual lenses, recognizing the battles that rage beyond the physical realm. Just as soldiers must be vigilant and alert on the battlefield, we, too, must be vigilant in our spiritual walk, always mindful of the forces around us. Our allegiance to God is paramount, shaping our decisions and priorities to reflect our commitment to His mission.

Prophetic Identity

I received my Spiritual Warfare Basic Training Camp (SWBTC) in Germany. While attending my first school of the prophets in Frankfurt, Germany, God revealed to me that I was called to be a prophet. We lived in Butzbach, Germany, a small German community with a rich history across from Rhein Bach Prison. At the time, I had been a Christian for about three years, and while the church was moving from spiritual warfare into the "name it and claim it" Era, God was preparing me for my destiny. Living in a region with a network of witches, I encountered demonic activity frequently, which God used to strengthen my prophetic walk. I attended the School of the Prophets in Frankfurt, Germany, under the guidance of Bob Godwin, not knowing the impact it would have on my life years later. Today, the foundation laid in Germany has led me straight into my Prophetic Spiritual Identity!

In this chapter, we must recognize how we all perceive or define our version of identity, whether it is saved or not. As believers, we must realize that our primary and only identity is in God, and the redeeming power of the Blood of Jesus is the only thing we can rely on in times of battle. We often create monuments of our achievements, but when challenges arise, these accomplishments cannot help us in any way. The identity God gives us is the only one that truly matters and can withstand even the most significant challenges. Discovering and embracing our spiritual identity is a profound and transformative experience. It shapes how we see ourselves and interact with the world around us. The spiritual warfare training we received in Germany was a pivotal moment that set the stage for a deeper understanding of our purpose and calling. By encountering demonic activity and facing challenges head-on, we are refined and prepared for the path ahead.

As we delve into the depths of our spiritual identity, we uncover layers of our being that go beyond mere physical existence. It is a journey of self-discovery guided by divine intervention and revelation. The School of the Prophets in Frankfurt was a beacon of light, illuminating the way to a higher spiritual plane. Under the mentorship of Bob Godwin, seeds of wisdom and insight were planted, destined to bloom in due time.

Our identity in God transcends worldly definitions and limitations. It is a source of strength and resilience in the face of adversity. Just as the foundation laid in Germany paved the way for a prophetic spiritual identity, so does our unwavering faith in God fortify us for the challenges ahead. Embracing our spiritual identity is not just a choice but a calling—a divine purpose woven into the fabric of our being.

My family and I faced a challenging situation that we felt ill-equipped to handle. However, I realized that my emotions and worldly desires were not what I needed to rely on in this battle. Just like a soldier in the military, I had to trust that God had prepared me for this fight. I may have yet to learn all the details of the mission, but I understood that we were confronting a challenge related to German warfare that required a spiritual approach. So, we took the first step forward.

5

COMPONENTS OF WAR

Now that we have delved into the concept of Germ Warfare, it is intriguing to explore its integration into our daily lexicon. "ART" connotes employing a specific skill set or a strategic approach. Similarly, warfare itself is often characterized by strategic maneuvers and calculated actions. Reflecting on the divine perspective, we find God, the ultimate creator, fashioned the world and pronounced it good. Yet, within this goodness exists a recognition of the harsh realities and challenges that permeate our existence. As inhabitants of this earthly realm, we are tasked with representing Him and carrying out our duties diligently until His awaited return, a notion underscored in the teachings of Luke 19:13. This profound responsibility underscores the importance of aligning our actions with divine principles and conducting ourselves with integrity and purpose as we navigate the complexities of life.

LUKE 19:13

. . .

"And he called his ten servants, delivered them ten pounds, and said unto them, Occupy till I come."

In this context, "occupy" has essential implications, particularly regarding banking activities. It suggests a sense of responsibility and active involvement in managing financial matters with care and expertise. Just as we trust in God to guide us and provide opportunities for growth and success, we also must nurture and develop the investments and talents entrusted to us.

This concept extends beyond simple financial transactions; it encompasses a broader understanding of the dynamics of war and warfare. By gaining knowledge and insight into these complex processes, we equip ourselves with the tools to navigate challenges and conflicts effectively. Therefore, "occupying" goes beyond merely conducting business; it delves into stewardship, empowerment, and strategic thinking, urging us to be proactive in our endeavors and mindful of the resources at our disposal.

The Enemy's Method of Approach: The Unseen Onslaught

Germ Warfare *is a strategy or weapon that intentionally uses germs or pathogens to weaken an enemy or opponent.* Its effectiveness is rooted not in the act of contamination but in how it is executed, making it akin to a "*secret weapon.*" This form of warfare can be likened to a siege—a tactic employed in warfare since ancient times. A siege entails surrounding an enemy fortress or stronghold, applying various pressures and strategies over time to weaken the adversary before launching a sudden and final attack to conquer them.

Historically, sieges could last anywhere from one to fifteen years. While modern terminology may only sometimes reflect this concept, acknowledging it is essential. Our society's focus on instant gratification often leads us to overlook the power of time. Though strength can arise from might and force, we must also appreciate the significance of patience and perseverance.

For instance, the overthrow of Saddam Hussein's regime took twelve years. It was a slow and gradual process. Similarly, malevolent forces use time as a weapon to undermine our futures and destinies through various tactics. These opposing forces have existed since the beginning of time and have had ample opportunity to observe human behaviors, patterns, and influences to develop effective strategies. Monitoring these influences is crucial; we will revisit this concept in future discussions. Although the average human lives in good health for about seventy-five to ninety years, time has existed far longer than humanity. These malevolent forces have been around for a long time and have refined their strategies over generations.

As a concept, time plays a significant role in shaping events and outcomes. It is often underestimated in a society that craves immediate results. The slow and steady progression of time acts as a silent yet powerful force, influencing the course of history and individual lives alike. Just as a siege methodically weakens its target over time, the passage of time can erode even the strongest fortresses of determination and resolve.

Furthermore, the intricate interplay between time and strategy extends beyond physical warfare and political conflicts; it also encompasses personal growth and spiritual battles. Unseen forces, like the pathogens in germ warfare, operate subtly and strategically over time to undermine our progress and potential. Recognizing this, we must embrace the wisdom of patience and resilience to withstand the relentless challenges of time and adversities.

. . .

THE ENEMY HAS FOUND a way to outlast us! Time can be a source of both pleasant surprises and disappointment. A siege is a strategic maneuver that relies on time to delay achieving a goal. It allows for extending the duration needed to reach the desired outcome.

IN 1 SAMUEL 30, David and his men arrived at Ziklag only to find that it had been destroyed by fire and their wives and children had been abducted. How did this happen? The Amalekites, a tyrant for generations, used time to their advantage, treating it as a secret weapon. Whether we consider this a siege or not, it is clear that the Amalekites' deep-seated hatred played a significant role in their strategy.

HE SAID that because the LORD hath sworn that the LORD will have war with Amalek from generation to generation ***(Exodus 17:1-16 KJV)***

GERM WARFARE IS a complex and intricate system designed with meticulous precision to operate independently, crossing boundaries effortlessly and replicating with a virulent efficiency comparable to the rapid reproduction of the letter "A." It targets the essence of human existence and the innate will to survive. Although Germ Warfare may seem like a singular weapon, it is critical for achieving devastating accuracy in its destructive objectives. Our adversaries strategically deploy German Warfare, employing many cunning strategies to sabotage individual pursuits and broader military campaigns. Unlike conventional Warfare, German Warfare does not merely hinder progress; its primary goal is annihilation.

THIS RAISES FUNDAMENTAL QUESTIONS:

. . .

- What are the defining characteristics of German warfare? How can we identify its presence and distinguish it from other threats?

- In the face of such a formidable adversary, what strategies can we employ to counter and ultimately defeat German warfare? What measures should be taken to neutralize its lethal impact?

- Why does germ warfare persist in targeting humanity? What drives this relentless assault on life, and how can we understand its motives to combat its destructive intentions effectively?

- Who truly stands as the enemy in the context of germ warfare? Is it a tangible foe or an abstract concept, and how do we navigate this complex landscape to safeguard our existence?

War Components:

Recalling God's message that "Germ Warfare is killing my identity," I began to grasp the naked truth of my prophetic walk and how to operate within the prophetic authority passed down through generations. As God's mouthpiece, the requirements are simple: do not question; obey! However, should we take the inner workings of German warfare more seriously?

Here are some key concepts to consider:

- **The Carrier:** The vessel or instrument used in the warfare.

- **Weapon:** The device or instrument employed in the fight.

- **The Enemy/Opposition:** Those who oppose or fight against another.

- **Target:** The destination or purpose aimed for.

- **Strategy:** The method being employed.

. . .

Reflecting on the profound message about German warfare, it becomes clear that this concept extends beyond physical combat and delves into spiritual warfare and the battle for one's identity. Understanding the intricacies of this warfare is essential for those operating in prophetic authority, as it illuminates the enemy's tactics to undermine our identity.

"*The Carrier*" takes on a more profound significance when viewed through spiritual warfare. It symbolizes the instruments used to carry out attacks and individuals who unknowingly spread negativity and falsehoods. Recognizing these carriers is vital to combating the destructive effects of Germ Warfare on one's identity and purpose.

Similarly, examining the concept of ***"Weapon"*** within spiritual warfare reveals a broader perspective. It expands beyond physical devices to include spiritual weapons such as prayer, faith, and discernment. By understanding the power of these weapons, one can effectively counter the lies and deceptions propagated by the enemy in the battle for identity.

"The *Enemy/Opposition*" gains deeper meaning when considering the spiritual forces in German Warfare. It represents those who directly oppose us and the spiritual entities working behind the scenes to sow discord and confusion. Recognizing and standing firm against this opposition is crucial for maintaining one's identity and prophetic walk.

Defining the "*Target*" becomes essential as one navigates the complexities of German warfare. This target is not merely a destination but embodies a purpose or calling the enemy seeks to derail. Understanding and aligning with this target is crucial in resisting attacks on one's identity and fulfilling the prophetic mandate bestowed upon us.

LASTLY, EXPLORING THE "*STRATEGY*" employed in German Warfare reveals the enemy's methods to undermine our prophetic walk. By discerning these strategies and aligning with God's divine plan, we can effectively combat attacks on our identity and walk in the prophetic authority inherited from generations past.

GERM WARFARE IS an unseen enemy or weapon that can disrupt people's destinies. While some argue that God would not allow dreams and destinies to be abandoned, the reality is that many individuals choose to give up on their dreams and ambitions. External factors often influence people's free will. The Apostle Paul teaches us that our struggle is not against flesh and blood but against principalities, powers, the rulers of darkness, and spiritual wickedness in high places (Ephesians 6:12).

Understanding the W.I.N System Headquarters

Opposition is a constant companion in the intricate tapestry of human existence, presenting itself as a formidable challenge that tests the very essence of our beliefs and resilience. For believers, this opposition often manifests in various forms, from external forces that obstruct our path and thwart our progress to internal struggles deeply rooted within our familial history. Yet, amidst the trials and tribulations, we find solace in the unwavering belief that we are imbued with the strength and fortitude to overcome any obstacle that dares to stand in our way. Through this steadfast faith and trust in a higher power, we discover the means to transcend our limitations and rise above adversity. As we journey through the labyrinth of life, it is imperative to recognize the transformative power of allowing God to enrich our spiritual arsenal and broaden our comprehension of the W.I.N. system. This divine framework empowers us to navigate the complexities of our Christian faith with unwavering confidence and trust.

Let's Read Eph. 6:12:

"For we wrestle not against flesh and blood, but against principalities, against powers, against the rulers of the darkness of this world, against spiritual wickedness in high places."

Paul, a former Roman soldier, brings a unique perspective to his writings as he often draws parallels between the strategies employed by his former enemies and the supernatural realm. This juxtaposition adds layers of complexity and intrigue to his teachings. By delving deeper into the term "Principalities," we uncover a rich tapestry of meanings and implications. The word originates from the Greek term "ar-khay'," which denotes not just a singular concept but a multifaceted one. It signifies a commencement, a chief, or a leader, encompassing various dimensions of order, time, place, and rank. Within the context of Paul's writings, "Principalities" can connote the idea of a foundational element, a corner piece in the grand scheme of things, a position of authority such as a magistrate, a source of power, a ruling entity, or a fundamental principle guiding actions. By grasping the breadth of this definition, readers can gain a more nuanced understanding of the intricate web of forces at play in the spiritual realm and how these dynamics influence the narrative of Paul's teachings.

Principality Summation:

. . .

1) BEGINNING, origin

2) the person or thing that commences, the "first person or thing in a series, the leader."

3) that by which anything begins to be, the origin, the active cause

4) the extremity of a thing

4a) of the corners of a sail

5) the first place, principality, rule, magistracy

5a) of angels and demons

NUMBER two (*under Principality Summation*) refers to the *"first person"* or thing in a series, the leader." Paul's teachings suggest that this is not just a single demonic force but rather the first in a series of generals. Many years ago, God revealed an old computer during my training in Germany. I still remember the rough edges that needed to be smoothed along the sides. Each document on that computer contained a list of demons, their assignments, locations, strengths, and weaknesses.

REFLECTING on my time at American Military University (AMU), where I graduated with a degree in international relations, I often ponder the complexities of global governance. It's fascinating to delve into the term 'principality' and its connotations. While traditionally associated with spiritual hierarchies, the term also describes small monarchies, highlighting the diverse nature of governance structures worldwide. The geopolitical landscape, comprising 195 countries and principalities, is a tightly woven mosaic of political systems and power dynamics. Among these, 193 nations are united under the umbrella of the United Nations, showcasing the profound interconnectedness of our world.

EXPLORING the theological implications of Ephesians 6:12 reveals a rich tapestry of interpretations across different religious traditions.

While theological discussions may vary, the overarching theme of spiritual warfare and the existence of unseen forces resonates across various beliefs. This concept of spiritual warfare mirrors the strategic maneuvers observed in global politics. Just as nations engage in intricate power plays and form alliances, the spiritual realm operates with similar calculated intent, often hidden from immediate perception.

THE W.I.N NETWORK is a complex system that operates with hostility, presenting itself as an individual grid or camouflaging as a collective strategy. Its primary goal is to infiltrate across generations, obstructing the realization of human aspirations and fulfilling God-given purposes, as highlighted in Psalms 139:16.

WHILE SOME MAY DISMISS this notion as absurd or fantastical, its tangible effects can be observed through the pervasive feeling among many individuals that their chances have slipped away or that their deepest desires will remain unattainable.

THE IMPACT of Germ Warfare on individuals is profound and complex. It involves two primary aspects:

1. **External** refers to internal conflict's outward effects or consequences. It represents the visible signs of the struggles individuals face within themselves. External effects include missed opportunities, stalled progress, and persistent stagnation or unfulfilled potential.

2. **Internal:** In contrast, this dimension explores the inner workings of a person's psyche. It encompasses a constant assault on one's self-worth, confidence, and overall mental well-being. The internal

struggle created by the W.I.N Network can lead to self-doubt, insecurity, and a distorted perception of one's abilities and values.

By understanding the interplay between these external and internal dynamics, we can begin to recognize the insidious nature of the W.I.N Network and its damaging impact on individuals. This understanding highlights the importance of identifying and combating these opposing forces that undermine our dreams and potential, encouraging us to reclaim our agency and work towards realizing our true aspirations.

The enemy's most effective tactic is to harm people from within. As we previously discussed the Ecological Theory and the various environmental factors that affect us daily, it's evident that these factors can invade our internal and external selves. In other words, trials, adversity, and setbacks affect not only one area but all aspects of our lives. It's important to remember that Germ Warfare is a deliberate and invisible attack that targets our minds and emotions, leaving a profound impact that can paralyze us in all areas of our lives. Let's reintroduce our components of war, provide their definition, and apply these components to a sample scenario for illustrative purposes.

- **The Carrier**: The vessel, package, or instrument used to transport harmful elements, just like how negative thoughts and emotions can be carriers of destructive impacts on our well-being and relationships.
- **Weapon:** The device used, akin to how harmful behaviors and toxic patterns can be weapons that harm ourselves and others on the battlefield of life.

- **The Enemy/Opposition:** One who opposes an enemy, representing the internal and external forces that challenge our well-being and mental health.
- **Target:** The destination aimed for, symbolizing our goals, dreams, and aspirations that the warfare within us can impact.
- **Strategy:** The method highlights the importance of developing effective coping strategies and resilience to combat the invisible attacks on our minds and emotions.

IDENTIFY WAR COMPONENT.

A FRIEND of mine invited me to go to the mall with her. While we were there, she spotted a group of girls she knew didn't like her. To avoid any confrontation, we decided to change our direction. However, those girls kept following us around the mall. Finally, as we were leaving, one of the girls looked at us and said, "I knew you were scared; that's why you kept walking." My friend chose to remain silent, but it was clear that this girl intended to keep causing trouble.

OKAY, let's test your application and discernment. In the space below, fill in what you believe is the weapon, carrier, etc., from the scenario. Hint: A person can never be the enemy; that position is reserved for Eph. 6:12.

THE CARRIER:

Weapon:
The Enemy/Opposition:
Target:
Strategy:

. . .

In our School of Prophets, we use this (above) framework to teach spiritual warfare to people of all backgrounds. Often, people view their opponent as the enemy but fail to recognize that they could be carrying the problem. When dealing with leadership, many tend to perceive the leader as the source of the issue but may hesitate to acknowledge them as a vassal or carrier.

Answers & Explanations:

A friend of mine invited me to accompany her to the mall. While we were there, she spotted a group of girls who she knew didn't like her. To avoid any confrontation, we decided to change our direction. However, those girls continued to follow us around the mall. Finally, as we were leaving, one of the girls looked at us and said, "I knew you were scared; that's why you kept walking." My friend chose to remain silent, but it was clear that this girl would continue to cause trouble.

Who is the Enemy? Satan, Demon, Imps according to Ephesian 6:12

We need to educate people on the importance of not attacking individuals. After all, a person is merely a vessel or a pawn influenced by someone else. To learn effectively, we must evaluate the entire situation by assessing the battlefield we are in. In this context, a mall can be considered a battlefield. Why? Because it is a place where the enemy chooses to strike. Whether the threat is in your territory or you are in theirs, the battle begins the moment confrontation occurs. Remember, understanding the battlefield is crucial in any war or conflict.

Who/What is the carrier (instrument) or vessel being used?

. . .

The carrier (vessel, instrument): The devil uses friends as instruments to bring conflict.

What is the weapon used? *Intimidation: Prey upon a weaker vessel to terrorize and to strong arm.*

Depending on what the enemy seeks to achieve, weapons can vary. The enemy always aims to destroy. In I Peter 5:8-9, it is written: "Be self-controlled and alert. Your enemy, the devil, prowls around like a roaring lion, looking for someone to devour. Resist him, standing firm in the faith..."

What is the target? Their peace of mind!

The situation can differ and even become more serious. The initial test usually checks if someone falls for it; if they do, the enemy proceeds with their next plan. However, in this case, the person did not fall for it and remained loyal to their principles. Although the enemy wanted to put their mind at ease, taking the bait could have resulted in physical harm.

What is or was the strategy? Unclear results due to limited information.

To recap, the enemy has a specific goal or agenda in mind. However, this conflict ended because they avoided falling into our trap. It's important to note that while this round is over and the young teens have emerged victorious, the enemy may devise another plan.

6

HOW THE COMMAND CENTER WORKS

Everything before this chapter serves as the foundation, the building blocks upon which more complex knowledge and skills will be added. Just as soldiers in Basic Training must continuously upgrade their training to meet annual requirements, so must we be willing to evolve and adapt to our challenges. The mock training exercises mentioned are a way to test our retention of new information and our ability to apply it effectively.

PAUL'S REFERENCE to Ephesians 6:12 calls for spiritual awareness and understanding of the spiritual battlefield surrounding us. By acknowledging the hierarchy of this battlefield, we can better equip ourselves to face the challenges that come our way. It is important to note that discussing the enemy's tactics does not glorify him, just as studying finance does not equate to worshiping money. Instead, it is about gaining knowledge and skills to navigate the complexities of life.

. . .

THIS BOOK SERVES as a guide to help us navigate the trials and tribulations of life. Being unprepared for the spiritual warfare we face can leave us vulnerable to the adversary's attacks. In contrast, with proper preparation and a deep understanding of the battlefield, we can stand firm against the onslaught of challenges that come our way. The unrelenting nature of warfare requires us to be vigilant and well-prepared, but with the proper knowledge and skills, we can face our adversaries with confidence and resilience.

Branches of Services

We often discuss the Bible but fail to examine its contents in depth. The Bible recounts the expulsion of Satan from heaven, along with a third of the angels who followed him. According to the scriptures, they conspired at Mount Hermon before descending to earth, where they defiled it. It is crucial not to overlook the order in their actions. Despite the Bible's assertion that Satan's kingdom remains united, we tend to divide it unnecessarily. This lack of understanding plays into the hands of our adversary. Formulating doctrines and myths that undermine the authority of believers only serves to create division among us rather than against him. Contrary to common belief, Satan's bid for heaven was a deliberate and calculated move.

THE DEPARTMENT OF DEFENSE was established with a clear objective in mind. Each department member comprehends their role, responsibilities, and protocols for smooth operations. Given the diverse branches of the United States military, they must operate independently and collaboratively globally.

1. Army

The Army's primary function is to safeguard the nation through ground warfare, utilizing infantry, armor, artillery, and other weaponry. Within the Army are two Reserve units—the Army Reserves and the Army National Guard—comprising civilian personnel with military training ready to serve when required.

2. Navy

The Navy's mission is to secure victories, deter aggression, and uphold maritime freedom. It also operates a Reserve component.

3. Coast Guard

The Coast Guard's duties encompass shoreline patrols, oil spill containment and cleanup, and interception of illicit sea-borne drug trafficking. Similar to other military branches, it maintains its own Reserve unit.

4. Air Force

The Air Force's principal role is to safeguard U.S. interests through aerial and space operations.

5. Marine Corps

The Marine Corps seizes and fortifies beachheads to gain strategic advantages over adversaries.

6. The Space Force

The Space Force is responsible for training, equipping, and organizing personnel to safeguard U.S. and allied interests in space and bolster capabilities for joint operations.

Within the W.I.N Network, we have visually depicted the workings of germ warfare within an individual, emphasizing the need to understand the enemy internally and externally. As we delve deeper into this understanding, we must also enhance our comprehension of the battleground. How does Ephesians 6:12 function effectively? What was the nature of the collaboration between Satan and the Angels? How do the various branches of the military operate effectively? The key lies in establishing robust communication channels.

The Power of Logistics

The word "*logistics*" originates from the Greek adjective "*logistics*," meaning "*skilled in calculating*," and is connected to the Latin term "*logistics*." It derives from the Greek word "*logos*," which refers to principles of thought and action. In military contexts, logistics pertains to transferring personnel and materials from one location to another and maintaining those materials. It is a crucial component of military deployment that allows forces to support ongoing operations or respond effectively to emerging threats.

EFFECTIVE LOGISTICS COMMUNICATION is essential when all branches of the United States military collaborate for a common cause. Each branch must integrate its resources, technology, and diverse expertise, coordinating every aspect to enable real-time information exchange. Strong communication enhances efficiency, operational maneuvers, and precision engagement. But how does the military unify as a cohesive unit?

EFFECTIVE COMMUNICATION within the military involves more than just sending messages; it ensures that every unit and branch is aligned. This process utilizes various communication channels, such as secure networks, encrypted devices, and established protocols, to maintain a timely and accurate flow of information. Coordination is critical, as different branches may possess unique capabilities that must be seamlessly integrated to achieve mission success.

. . .

Additionally, military logistics encompasses a range of activities beyond transportation and supply chain management. It includes strategic planning, resource allocation, risk assessment, and contingency planning. The logistics function ensures that troops have the necessary equipment, ammunition, and provisions to carry out their missions effectively.

7

MILITARY MANEUVERING & COMMUNICATION

In the previous chapter, we delved into the intricate details of the W.I.N System and its comparison to Ephesians 6:12. Our exploration began by shedding light on German warfare. This concept reverberates through individuals' external and internal realms. However, it's imperative to acknowledge that German warfare transcends individual experiences, permeating every facet of society and turning it into a battleground. Paul's message to his contemporaries was clear - the people of God are not only navigating a hostile world. Still, they are also engaged in an enduring struggle against the demonic forces mentioned in Ephesians 6:12, a battle that will persist until the second coming of Christ.

THE FAILURE TO understand one's adversaries can lead to devastating consequences, as evidenced by historical data such as statistics from the Vietnam War. These figures paint a stark picture - with 61% of young men aged 21 and younger accounting for the casualties, it becomes evident that inexperience and youth can significantly impact survival rates in times of conflict. The sobering reality of war looms over our generation, underscoring the critical importance of effective communication in navigating such challenges. Whether verbal or nonverbal, communication is pivotal in shaping strategies and fostering unity amidst chaos. A deeper exploration of this crucial aspect awaits as we progress through the upcoming chapters. Remember, in the tumult of war, communication can be a beacon of hope, guiding individuals through the darkest times and forging bonds that withstand the test of adversity

Types Of Strategies

A military strategy is a plan that outlines the operations and movements needed during a battle or war. Regardless of its length, a strategy's effectiveness is determined by its precision and accuracy. Paying attention to details and employing specific methods is crucial to accomplish a mission. Below are some military strategies and their definitions, illustrating the importance of a well-defined plan.

Remember: "*In warfare, strategy involves the comprehensive planning of campaigns before engagements, while tactics focus on the actual maneuvers and specific actions taken to achieve victory in a battle." --Encyclopedia Britannica*

1. Precision Engagement

"*Precision Engagement consists of a system of systems that enables joint forces to locate the objective or target, provide responsive command and control, generate the desired effect, assess the level of success, and retain the flexibility to reengage with precision when required.*"

. . .

STRATEGIC PLANNING, the crucial element for winning any war, requires more than one approach. It necessitates a multi-dimensional mindset to project a successful plan and strategy accurately. A military example of precision engagement systems is provided below:

- *CONTROL & Communication Stations*
  - *Satellites and Space Communications*
  - *Base Headquarters and Headships*
  - *Rules of Engagement*
  - *Boundaries and Borders*
  - *Ground Troops*
  - *Air Troops*
  - *Local Information Stations*
  - *Combat Sittings*
  - *Real-Time Intelligence*
  - *Media Feedback*
  - *Special Forces Instructions*
  - *General Strike Force*

EACH COMPONENT in the precision engagement system plays a vital role in ensuring success on the battlefield. Control and communication Stations act as the nerve center, facilitating seamless coordination between different units. Satellites and Space Communications provide crucial data. Integrating various components is essential for effective military operations, particularly in maintaining connectivity even in remote locations. Base Headquarters and Command Centers serve as strategic hubs, overseeing operations and deployments.

CLEAR RULES of Engagement and defined boundaries are crucial for maintaining order and preventing unnecessary escalation. Ground Troops and Air Troops form the backbone of the military force, executing missions with precision and agility. Local Information

Stations collect intelligence from the field, providing decision-makers with real-time insights.

Combat Settings and Real-Time Intelligence ensure that commanders are well-informed, allowing them to adapt swiftly to changing circumstances. Media Feedback and Special Forces Instructions help manage public perception and facilitate the effective execution of specialized operations. The General Strike Force is always prepared to deploy rapidly and decisively when a swift and robust response is necessary.

By integrating these components into a cohesive system, precision engagement maximizes the effectiveness of military operations, minimizes collateral damage, and enhances overall mission success rates.

Each division or section must work together in a war zone and communicate all vital information to ensure success. This Communication helps prevent friendly fire and ensures that all parties know each other's positions. However, delays and casualties can occur when information is not shared appropriately. Precision engagement is achieved when all parties work together effectively and accurately to hit their designated targets.

Upon analyzing the entities mentioned, we can observe that the first one is Communication. Significantly, this Communication is not limited to a specific dimension; it extends upwards and downwards through each entity. One form of Communication covers the entire landscape by land and sea, while another involves space. With the help of satellites, we can track global gatherings, conversations, and the planning of diverse terrorist attacks worldwide. Although it is not possible to intercept every warning, we can at least position ourselves to stop most of them. Therefore, it is crucial to establish Communication with leaders worldwide through various means—land, sea, and space—to combat the threat of terrorism effectively.

As we aim to improve our Communication, we must consider the next phase, which involves establishing headquarters and leadership locations. We need to determine the appropriate number of headquarters and leadership positions required. In the military, having a plan in place before action is necessary. We cannot afford to wait for a war to break out before planning; we must be proactive and think ahead.

We have had to develop new ideas to address the catastrophe at hand, leading us into a new era of warfare. We understand that the President is the only Commander-in-Chief, and there can only be one. While many commanders exist, only the Commander-in-Chief has the authority to issue orders for war. The Commander-in-Chief has a team of generals and commanders who provide various war objectives, extraordinary intelligence, and strategies to help win battles. These commanders work behind the scenes to explore alternative solutions under the Commander-in-Chief's direction.

Let us take a moment to apply what we've learned so far. Opposition often acts as the Commander-in-Chief within its sphere of influence, delegating smaller divisions, such as Germ Warfare and other strategies, to derail you from achieving your goals. To effectively counteract the opposition, it is essential to recognize it and identify its tactics and weapons.

1. **Dominant Maneuver**

. . .

This strategy involves the multidimensional application of information, engagement, and mobility capabilities to position and employ dispersed joint air, land, sea, and space forces to accomplish assigned operational tasks.

Dominant maneuvers are a crucial strategy used in both land-based and international contexts. This tactic consists of two key components: strategic and operational. The strategic aspect encompasses the overall approach or methodology employed, while the operational facet focuses on the specific mechanisms and systems implemented.

At tiie iieart of undcrstanding this tactic is "dominancc." Λ lcader must identify the most effective means to assert dominance over an adversary or situation. This requires thorough analysis, asking relevant questions, and preparing to deploy the necessary forces, troops, and weaponry to achieve a swift victory.

It's essential to recognize that dominance, while not always equating to effectiveness, can be one of the most successful strategies when used wisely. By emphasizing the potential success of dominance, leaders can guide their forces toward a position of strength, enhancing their chances of success and instilling a sense of hope and optimism.

This strategy highlights the significance of proactive decision-making, strategic planning, and calculated risk-taking to achieve dominance on the battlefield or in diplomatic negotiations. Mastering the art of dominant maneuvers, with a strong emphasis on strategic planning, instills a sense of preparedness and confidence in the audience.

I. Operational Maneuver

An operational maneuver is a military strategy involving decisive actions to compel the enemy to submit to one's will. In combat, significant operations are marked by victorious outcomes in battles or engagements. The concept of operational maneuver can be understood as a collection of methods aimed at bringing the opposition to a complete defeat. It is essential to comprehend how these methods work together to establish a dominant strategy to defeat or weaken the enemy.

Our primary goal is to enhance our capabilities by gathering extensive information and mastering precision engagement strategies. Precision engagement is a multifaceted approach that relies on pinpoint accuracy, unwavering determination, and a comprehensive skill set to effectively neutralize adversaries' weaponry and tactics. We can take proactive measures by developing these essential skills instead of reacting defensively. This proactive stance allows us to embody the essence of authentic precision engagement, ensuring that our operations are successful and strategically advantageous.

Back To The W.I.N System

The W.I.N. System is a practical approach that illustrates different methods used in a process. In a previous discussion, we discussed Germ Warfare, which can manifest in external and internal forms. Internal Germ Warfare focuses on a person's self-worth, confidence, and other internal aspects. Regardless of their situation in life, Germ Warfare can quietly undermine the human spirit. The Formational Siege section in this headquarters addresses this issue.

. . .

SIEGES COME IN TWO CATEGORIES:

(1) **Generational**
(2) **Territorial.**

IN OUR PREVIOUS DISCUSSION, we explored various methods and tactics employed within a system, including Operational Maneuvers, Precision Engagement, and Dominant Maneuvers. These tactics are essential components of the system, each with unique characteristics. When we refer to generational and territorial warfare, we are explicitly talking about a type of warfare known as sieges. A siege involves surrounding enemy forces and waiting for them to surrender. A Formational Siege is a specific approach to siege warfare where the attacking forces use a distinct formation to encircle the enemy and capture their territory.

Generational Siege

Generational Siege is a term that describes a profound and ongoing struggle that affects a family over multiple generations, influencing various aspects of their lives, including lineage, descendants, and financial stability. This struggle can persist across generations, creating a significant impact. For instance, if a family member, such as an aunt or grandmother, struggles with alcoholism, this pattern may continue to affect not only immediate family members but also future generations. It's also important to recognize that **Generational Inheritance** can include positive and enriching legacies. These inheritances may be tangible assets passed down through an estate or a renowned family lineage known for producing innovators or visionaries. Such positive inheritances create a legacy that is handed down through successive generations.

. . .

GENERATIONAL INHERITANCE IS a broad concept that encompasses more than just financial assets. It can also refer to blessings, talents, gifts, and other valuable attributes passed from generation to generation. These non-material inheritances contribute to a continuum of prosperity and fulfillment within the family lineage, shaping the family's identity and future.

Territorial Sieges

Territorial Siege explores the complex dynamics of land ownership and the impact of one's voice. A crucial aspect highlighted in the accompanying chart is the relationship between "Your Voice," authority, and position. Examining the historical context that informs this strategic approach is essential to understand this correlation fully. Throughout history, religious conflicts have been significant drivers of warfare, and even today, political discord often catalyzes armed conflict. However, religious differences primarily fueled past conflicts. Interestingly, these conflicts were ideological and centered around territorial expansion. While land ownership remains a critical factor in social structures today, contemporary regulations govern these territorial disputes with greater sophistication.

SEIZING territory was more than just a conquest; it represented dominion over the acquired lands. Legendary figures like Alexander the Great are remembered for their military prowess and strategic skills. Capturing enemy territories was frequently accomplished through the tactic of laying siege. Armies would besiege their adversaries, employing strategies of attrition and gradual dismantling. In historical contexts, when a new ruler aimed to claim land, the ousted ruler faced a stark choice: surrender or confront severe consequences, including torture and death. Regardless of the outcome, the new ruler would assert control over the land and its people. Sometimes, the deposed ruler would be retained as a vassal, an intermediary between the new ruler and the subjects to facilitate control. However, defiance from the former ruler often led to brutal repercussions. In all scenarios, the new ruler subjugated the land and its inhabitants.

For believers, it is vital to understand the multifaceted nature of spiritual warfare, as referenced in Ephesians 6:12. This encompasses a variety of strategies, sieges, and maneuvers. Such awareness helps identify the battles we encounter and guides a thoughtful response. Instead of reacting impulsively, a proactive stance is advisable, recognizing the complexity of the spiritual battlefield and the ongoing need for strategic enhancements to achieve victory.

8

SOUND IS A WEAPON OF MASS DESTRUCTION (WMD)

Biological warfare, or Germ Warfare, utilizes natural toxins or infectious agents like bacteria, viruses, insects, and fungi to cause harm or incapacitate living organisms during conflicts. The use of biological weapons is a highly controversial and dangerous practice due to the devastating potential impact on both military personnel and civilian populations. Unlike conventional weapons, biological weapons can be complex to detect and can have long-lasting effects on the environment and public health. Biological warfare is considered a distinct category of weapons of mass destruction (WMDs) due to its unique characteristics and the challenges it poses in terms of defense and containment.

The deployment of biological weapons can lead to widespread casualties, disruption of essential services, and long-term health consequences for survivors. The threat of biological warfare has prompted the development of international treaties and agreements aimed at preventing the production and use of such weapons. While the focus of this discussion is on biological warfare, it is essential to note that other emerging technologies and tactics are shaping the future of warfare. One such innovation is using sound as a weapon, which has been increasingly recognized for its potential in military operations. Sound can disrupt enemy communications, disorient adversaries, or cause physical harm. The strategic use of sound as a weapon represents a new frontier in military technology and underscores the ever-evolving nature of conflict in the modern world.

LET'S Read 1 Cor.14:6-8

"THINK, friends: If I come to you and all I do is pray privately to God in a way only he can understand, what are you going to get out of that? If I don't address you plainly with some insight or truth or proclamation or teaching, what help am I to you? If musical instruments—flutes, say, or harps—aren't played so that each note is distinct and in tune, how will anyone be able to catch the melody (sound) and enjoy the music? If the trumpet call can't be distinguished (heard), will anyone show up for the battle?"

THE FIRST VERSE of 1 Corinthians 14 is similar to the last verse of chapter 12. Both verses encourage believers to seek spiritual gifts, especially the gift of prophecy, while pursuing love. Paul repeats the phrase "eagerly desire" about the gifts in both 1 Corinthians 12:31 and 14:1, highlighting the importance of these gifts to the church. By repeating this message in these two chapters, Paul emphasizes that love, as described in chapter 13, must be the foundation for using spiritual gifts in the church.

PAUL ENCOURAGES us to pursue love diligently. The word "pursue" used here is quite strong, as Paul himself uses it elsewhere to describe his former intense pursuit of the church as a persecutor. For instance, he says, "I persecuted the church" (1 Cor. 15:9). Therefore, we must pursue love as relentlessly as Paul pursued the church. Paul often uses this word as a metaphor for spiritual effort, as seen in other places like Rome. 9:30, 31; 12:13; 14:19; Phil. 3:12, 14; 1 Thess. 5:15; 1 Tim. 6:11; 2 Tim. 2:22.

My focus is **on verses 6-8,**

If musical instruments—flutes, say, or harps—aren't played so that each note is distinct and in tune, how can anyone catch the melody (sound) and enjoy the music? If the trumpet call can't be distinguished (heard), will anyone show up for the battle?"

Musical instruments such as flutes or harps require a skilled touch to coax each note with precision and harmony, ensuring the sound is rich and melodious. The essence of sound lies in its energy, which propagates through a medium like air, carrying the vibrations that create the audible sensation perceived by our ears. These vibrations, when consistent, give rise to pleasing sounds, while irregular ones result in discordant noise that can be jarring to the senses. It is fascinating to note that various factors, including a person's emotional state, the surrounding environment, or the intrinsic qualities of an object, can influence these vibrations, creating a unique auditory experience that resonates with those who perceive it.

Sound, whether music or speech, always serves a purpose, intentional or otherwise. In warfare, deploying Sonic and Ultrasonic weapons represents a chilling example of sound being weaponized to harm or incapacitate adversaries deliberately. The use of sound waves to induce pain, disorientation, or discomfort underscores the potency of sound as a tool for aggression and control.

. . .

However, it is crucial to recognize the potential dangers of excessive exposure to loud sounds, as they can harm our auditory system. Prolonged exposure to intense sound levels can lead to various issues, including hearing loss and damage to delicate structures within the ear, like the eardrum and cochlea. The fragility of our hearing mechanism highlights the importance of safeguarding our ears from harmful noise levels and practicing responsible listening habits.

The biblical reference in 1 Corinthians 13:1 parallels the significance of the words' power' and' love.' It underscores that words spoken without love are akin to the harsh clangor of noisy instruments devoid of meaning and warmth. The imagery of "sounding brass" and "a tinkling cymbal" is a metaphor for communication that lacks empathy and compassion, resonating discord rather than harmony.

In essence, the passage reminds us of the profound impact our words can have, likening them to potent weapons capable of inflicting emotional harm or destruction. By exercising care and mindfulness in our speech, we can harness the transformative potential of language to uplift, inspire, and bring healing rather than cause harm or discord. When imbued with love and empathy, our words can build bridges, foster understanding, and nurture relationships, ultimately shaping the world for the better.

The frequency of The Wind Of God

As children of God and especially as leaders, we must recognize the devastating impact that hurtful words can have. Words possess significant power; when directed at someone, they can create a destructive force akin to a strong wind. These harmful vibrations can unleash negative energies that can incapacitate a person, damaging their emotional well-being.

. . .

In the Gospel of John, Jesus illustrated the work of the Holy Spirit by comparing it to the wind during his conversation with Nicodemus, a religious leader. Like the wind, he said you can hear the Spirit but cannot determine its origin or direction. Similarly, explaining how people are born of the Spirit is difficult. Just as you can listen to the wind without knowing where it comes from or where it is going, you can't easily explain the workings of the Spirit in a person's life (John 3:8 NLT).

The comparison between the Spirit of God and the wind is significant because the exact words in Hebrew and Greek can refer to wind, spirit, or breath, depending on the context. Wind or breath symbolizes life. In the Book of Ezekiel, there is a powerful depiction of bones coming together without any breath or life in them. Then, the Lord commands the four winds to breathe life into the lifeless body. This event illustrates the power of wind or breath, akin to the power of the Holy Spirit.

Wind consists of moving air molecules that create vibrations when passing through objects, such as trees. These vibrations are perceived as sound by our ears. Unlike music or a person's voice, the sound of the wind lacks a specific frequency or tone.

Thus, 2 Corinthians 10:5 instructs us to "*cast down imaginations, and every high thing that exalteth itself against the knowledge of God, and bringing into captivity every thought to the obedience of Christ.*" Understanding sound's power and discerning which frequencies should be avoided is essential.

What The Enemy Fears About Us

Humanity has been granted an incredible privilege and responsibility as representatives and partners of God in the world. Despite being made from dirt and mud, humans can be transformed and united with divine life and wisdom. They can navigate both the spiritual realm and the physical universe as if both are their native home. As physical beings created in the image of God, this unique aspect is something that no other creatures possess.

GIVEN the context in which most of the Bible was written, it is very likely that the devil could not accept the idea of a "lower creature" being invited into the love and family of God, especially when placed on an equal or even higher footing than more powerful spiritual beings. The animosity of the enemy towards God is evident and well-documented throughout scripture. The devil desired to be God but ultimately failed, and this fuels his hatred towards God. He knows that his power is weaker than God's and that God's authority restricts him. Consequently, he seeks to take out his frustration on humanity by lying, tempting, and blinding us.

THIS DIVINE CONNECTION between humanity and God forms a profound and sacred bond that transcends the limitations of our earthly existence. This relationship elevates us to a higher plane, allowing us to tap into our Creator's infinite wisdom and love. Despite our humble origins, we are granted the remarkable opportunity to partake in the divine nature and reflect God's image in our lives.

THE SPIRITUAL AND physical realms intertwine in a harmonious dance, enabling us to navigate gracefully and purposefully. This duality of our existence showcases the multifaceted nature of our being, as we are called to embody the divine essence while embracing our earthly roots. Achieving this delicate balance requires us to constantly seek alignment with God's will and the guidance of the Holy Spirit.

The devil's envy and resentment towards humanity stem from his prideful ambitions and ultimate downfall. His rebellion against God led to his eternal separation from the divine presence, filling him with bitterness and malice towards God's beloved creation. His deceptive tactics and relentless attacks on humanity are driven by a desire to undermine God's plan and divert us from our true purpose.

Despite the adversary's attempts to deceive and corrupt us, we are empowered by the light of God's truth and the strength of our faith. Through prayer, meditation, and spiritual discernment, we can shield ourselves from the enemy's snares and walk confidently in the path of righteousness. Our identity as God's children and heirs to His kingdom grants us the authority to resist the devil's schemes and stand firm in our convictions.

In the face of adversity and temptation, we must hold fast to our faith and trust God's providence. Our unity with the divine enables us to overcome all obstacles and emerge victorious in our spiritual battles. By embracing our role as God's representatives on earth, we fulfill our destiny and honor the sacred covenant that unites us with our Creator for eternity.

1. **The Enemy Fears Our Authority**

PAUL EMPHASIZES that the enemy operates at a level of strategy that we often overlook. He wants to illustrate that they function as a cohesive unit, leaving no vulnerabilities. Ephesians 6:12 describes a demonic hierarchy that plays various roles on the battlefield, and without proper understanding, we risk becoming casualties. Until now, we have tried to convey that we were born into a hostile environment known as Earth. However, despite humanity's fall, God never revoked the authority given to us in Genesis 1:26.

THEN GOD SAID, "Let us make mankind in our image, in our likeness, so that they may "rule" (dominion) over the fish in the sea and the birds in the sky, over the livestock and all the wild animals,[a] and over all the creatures that move along the ground.

1. **The Enemy Fear Our Presence**

WE ALSO SEE that God establishes the Noahic Covenant after the flood, which adds to our (believers') authority. Gen.9:1-3;

"THEN GOD BLESSED Noah and his sons, saying to them, "Be fruitful and increase in number and fill the Earth. 2 The fear and dread of you will fall on all the beasts of the Earth, and on all the birds in the sky, on every creature that moves along the ground, and on all the fish in the sea; they are given into your hands. 3 Everything that lives and moves about will be food for you. Just as I gave you the green plants, I now give you everything."

AT FIRST GLANCE, there appears to be no difference. However, the difference lies in verse two of 2 of Gen. 9

. . .

"The fear and dread of you will fall on all the beasts of the Earth, the birds in the sky, every creature that moves along the ground, and all the fish in the sea; they are given into your hands."

Genesis 1:28 and 9:3 state that our authority remains unchanged despite the flood. This enduring authority signifies the divine mandate given to humanity to rule over the Earth and its creatures. However, a shift occurred in the dynamics between humans and animals after the flood. They were no longer merely subservient; now, they possessed a sense of fear toward our presence. This change reflects a more profound spiritual truth: even in adversity, all creation recognizes our inherent power and dominion.

Although adversarial forces may try to undermine our confidence and create doubt about our abilities, they tremble before our true potential. Through diligent study of the Old Testament and the establishment of divine covenants, we can unravel the intricate tapestry of these spiritual principles and their profound connection to the redemptive work of Christ. These ancient teachings serve as a guidepost, illuminating our path toward a deeper understanding of our place in the cosmic order.

When our presence becomes a threat to the forces of darkness, our voice emerges as a powerful weapon, resonating with the authority granted to us by the Creator. We wield unparalleled influence in the spiritual realm by aligning our will with divine purpose. As we navigate life's challenges, let us remember that our true strength lies not in earthly might but in the unwavering faith that empowers us to speak truth and manifest change in the world around us.

9

VOICES ON THE BATTLEFIELD

The previous chapter explored the powerful impact of sound waves as a destructive force, emphasizing how sonic frequencies can damage delicate eardrums, potentially leading to hearing impairments and a disconnection from the divine voice. Building on this idea, the current chapter will further examine the intricate relationship between words and sound vibrations, illustrating how language travels with purpose and intent through sound waves.

As you step onto the battlefield, you must recognize that you are not navigating this terrain alone; your adversary is also strategizing. Understanding the nuances of warfare's language, including its constructive and destructive aspects, is essential. Without a solid grasp of this linguistic arsenal, you may find yourself at a disadvantage, vulnerable to the tactics employed by the opposition.

Words, rich with meaning and emotion, are the ammunition that drives communication during conflict. They can shape perceptions, influence actions, and ultimately determine the outcome of encounters. By acknowledging the strength of language as a tool of warfare, individuals can harness its power to navigate the complexities of battle with clarity and purpose.

Pro.26:2;

"As the bird by wandering, as the swallow by flying, the curse causeless shall not come."

The phrase "As the bird by wandering, as the swallow by flying, the curse causeless shall not come" highlights the idea that curses, like birds in flight, must have a cause. A curse is a powerful expression of negative emotions, often driven by anger or frustration, and it serves as a verbal way to inflict harm or misfortune on someone. When someone curses, they release energy into the universe that can impact both the target and the speaker. Cursing reflects a person's inner emotional state, as the spoken words carry significant weight and can linger in the environment, influencing the energy of those nearby.

Moreover, the power of words, including curses, lies in their ability to shape reality. Words are more than mere sounds; they embody intention and meaning. The vibrations created by spoken words can affect the energy fields of both the speaker and the listener, resulting in a ripple effect that extends beyond the moment of expression.

Thus, this quote reminds us of the interconnectedness of words, emotions, and actions. It suggests that every curse, like every action, has consequences and a source. Just as a bird's flight is purposeful, our words should be guided by mindfulness and intention, recognizing the potential impact they may have on ourselves and others.

According to Proverbs 26:2, the power of words is emphasized. Blasphemous words can inflict harm only when they resonate with the recipient's innermost feelings, such as harboring unforgiveness or lingering bitterness. This verse underscores the impact of words on the human psyche, suggesting that negative words can indeed wound only if they align with an individual's existing vulnerabilities.

Moreover, the concept of incantation is introduced, which defines it as the practice of using words as a form of magical spell or charm. This notion delves into the idea that words possess a mystical quality that can influence reality or invoke supernatural forces. By likening words to incantations, the text highlights the inherent power of language and how it can shape circumstances or outcomes.

Furthermore, mentioning words used as curses expands the idea of language as a potent tool. It illustrates the darker side of communication, showing how words spoken with malice or ill intent can manifest as curses that bring harm or misfortune to their target. This serves as a cautionary reminder of the responsibility accompanying verbal expression and the need for mindfulness in our speech.

In essence, Proverbs 26:2 and the discussions on incantations and curses underscore the profound influence of words on individuals and their surroundings. They prompt us to contemplate the significance of language, urging us to be mindful of the words we speak and the intentions behind them, as they have the power to build up or tear down, to bless, or to curse.

. . .

The Birthing Of Witchcraft

Ironically, the internet defines witchcraft as the use of magic specifically for evil purposes through spells. However, isn't this just the nature of words? This leads us to the conclusion that witchcraft is essentially the act of combining and merging words in agreement and then releasing them in unison. It needs to be made clear when witches first appeared in history. One of the earliest records of a witch can be found in the Bible, specifically in the book of 1 Samuel, believed to have been written between 931 B.C. and 721 B.C. This book tells the story of King Saul, who sought out the Witch of Endor to summon the spirit of the dead prophet Samuel for guidance. Traditionally, the term "witchcraft" refers to the use of magic or supernatural powers to inflict harm or misfortune on others, and this remains the most widespread interpretation today.

The original Hebrew word in Exodus, translated as "witch," is *mekhashepha*. Unfortunately, we can only rely on modern interpretations, as we cannot know what that word meant when Exodus was written thousands of years ago. The Ben Yehuda Hebrew Dictionary translates *mekhashepha* as "witch." In Arabic, this verb means "to reveal" or "to discover" and can also mean "to pray" or "to practice magic," literally translating to "to pronounce or mutter spells."

Some individuals intentionally harm others, and demonic forces may influence them. These individuals may use curses or other means to cause damage. Unfortunately, even spiritual leaders such as prophets or pastors can be influenced by opposing forces, potentially causing harm to their followers without realizing it. This can severely affect the faith and future of those impacted. The real question is whether these leaders know their actions or if something more sinister is at work.

. . .

Witchcraft has been a subject of fascination and fear throughout history, with societies often viewing practitioners with suspicion, leading to persecution. The practice of witchcraft varies widely across cultures and traditions. Some view it as a form of empowerment and a connection to nature, while others condemn it as dark and malevolent.

Throughout literature and folklore, witches have been portrayed in various ways, from the wicked old crone stirring a cauldron to the wise healer using herbs and incantations to help others. These depictions reflect witchcraft's complex and multifaceted nature, highlighting how it can encompass light and dark aspects. Despite the negative connotations often associated with witchcraft, many modern practitioners view it as a spiritual path centered on self-discovery, healing, and connection to the natural world. Wicca, for example, is a contemporary pagan religion that draws upon pre-Christian beliefs and practices, emphasizing harmony with nature and the earth's cycles.

Power Of Unity

In biblical times, the people of God initially lived under a theocracy. However, some angels chose to abandon their faith and turned towards paganism, pantheism, and polytheism. Through primary research, one can define and explain these various religious beliefs as follows:

- **Paganism** is an ancient religion that combines various streams of spirituality, philosophy, devotion, and ritual practices. It is syncretic and fluid, with undefined boundaries and subjective interpretations. This quality makes paganism a relatable form of religion for the postmodern age, much like in ancient Rome. It's important to emphasize that paganism is not inherently tied to concepts of morality.

- **Pantheism** is the belief that God and the universe are not separate entities but the same. In essence, this means that God is everything, and everything is God. The term "pantheism" derives from the Greek words "pan," meaning all, and "theos," meaning God. Its development can be traced back to the eighteenth century.
- **Polytheism**, in contrast, naturally evolves from animism, a primitive form of polytheistic belief. In polytheism, the deities associated with nature become distinct from their corresponding elements, such as mountains, trees, rivers, and animals. These entities transform into gods and goddesses in their own right, requiring worship and appeasement—not just as the deities of a specific body of water, for instance, but in a broader sense, as Poseidon, the God of the Sea. In polytheism, the myths, rituals, and religious practices aimed at understanding and appeasing these gods become increasingly intricate and elaborate.

THIS TOPIC IS QUITE broad and requires individual study. However, for now, let's focus on the immediate danger of being chased on the battlefield, having a hovering presence over your shoulder, and being apprehended as a prisoner of war. As the Apostle Paul describes in Ephesians 6:12, Demonic Declarations and Monitoring Spirits are the biological weapons being used.

LET'S READ GEN.11:4;

. . .

And they said, Go to, ***let us build*** *us a city and a tower, whose top may reach unto heaven; and let us make us a name, lest we be scattered abroad upon the face of the whole earth.*

In Gen.11:4, Nimrod has encouraged the people to come together and build a name for themselves. However, in Ge, 11:6, it reads;

"The LORD said, "If as one people speaking the same language they have begun to do this, then nothing they plan to do will be impossible for them."

Their unholy unity formed a powerful bond that transcended ordinary collaboration. This synergy drew strength from the depths of darkness, creating an unbreakable connection that fueled their ambitions. Their belief in invincibility was a fleeting thought and a deeply ingrained conviction that resonated with intensity. Each word they spoke seemed to carry its weight, pulsing with a sinister energy that wove together a grand design—a malevolent tapestry of intentions and desires. They saw themselves as masters of their destiny, architects of a future where nothing could stand in their way.

Their confidence bordered on arrogance, bolstered by unwavering faith in their collective power. To them, limitations were mere illusions, obstacles to be effortlessly overcome in their relentless pursuit of dominance. Yet, amid their fervor and zeal, a shadow of caution lingered—a whispered acknowledgment of a force greater than their own. The mention of God served as a sobering reminder of a higher authority, a potential barrier that could shatter their illusions of omnipotence. This was a subtle acknowledgment of a boundary they dared not cross despite their bravado.

. . .

Unified Witchcraft

We are exploring the critical concept of survival on the battlefield, a skill that requires a deep understanding of battlefield dynamics. It is essential to emphasize that the ability to interpret the nuances of the battlefield is a fundamental aspect of military strategy and tactics. This intricate skill involves recognizing the physical terrain and comprehending the psychological, strategic, and tactical elements.

The complexity of warfare goes beyond what any book or manual can fully cover regarding its components and intricacies. Leaders and allies must navigate a treacherous landscape, continually assessing threats and opportunities while remaining cautious of potential betrayals. The battlefield is not just a physical space but also a psychological and emotional arena where trust and loyalty are tested at every turn.

Furthermore, the references to witchcraft, spells, and incantations add a metaphorical layer to the discussion, highlighting the insidious nature of deceit and manipulation in conflict. Drawing parallels to historical figures like Nimrod underscores the enduring nature of deception and the risks of blindly following charismatic but misguided leaders.

This text underscores the importance of critical thinking, discernment, and moral clarity in navigating the complexities of warfare and human conflict. It is a cautionary tale against falling prey to falsehoods and manipulation, urging individuals to uphold integrity and ethical principles even in the most challenging and chaotic circumstances.

Vassal Voices Are Camouflaged

In a previous discussion, we explained the concept of a vassal in the Old Testament. Kings employed vassals to serve as intermediaries between the new ruler and the people of the land. The new king would leverage the old king's influence to assert control over the population. However, if the old king refused to surrender, he could face torture or execution. Ultimately, the new king would always claim ownership of the land and its inhabitants. Additionally, we discussed how to identify different components of war, focusing on key players and factors involved.

Here is a recap of the distinctions we made:

- **Carrier:** The vessel, package, or instrument used for transport.
- **Weapon:** The device or instrument employed in combat.
- **Enemy/Opposition:** The force that opposes an enemy; someone who fights against another.
- **Target:** The intended destination or objective.
- **Strategy:** The method employed to achieve a goal.

It is important to emphasize that the previous discussion overlooked a significant aspect regarding how adversaries might exploit leaders as vassals. It is crucial to distinguish between the term "vassal" and the word "vessel." The former indicates a shift in allegiance, while the latter typically refers to a container or instrument. While earlier scenarios focused on fundamental aspects of warfare, a deeper understanding is required as individuals strive to follow God's will and grow in faith. This raises the question: how does one discern vassals amidst the chaos of the battlefield?

. . .

Many religious institutions foster a culture of fear around authority figures, shielding leaders from accountability for their actions. This lack of transparency and integrity can lead to grave consequences, such as youth falling victim to sexual abuse and the rise of criminal gangs. Furthermore, it highlights the opportunistic behavior of politicians who seek to gain favor with voters by making appearances at churches, often prioritizing personal interest over genuine service.

Despite sharing the same battlefield, many individuals become victims of harm inflicted by those they trust. The concept of camouflage—commonly associated with hiding one's presence through disguise—has its parallel in religious settings, where falsehoods conceal the truth, friendly smiles mask deceit, and financial influence enforces silence.

Let us examine a controversial scripture of all time,

Ps. 105:15:

"Do not touch my anointed ones; do my prophets no harm."

The verse states, "Do not touch my anointed ones; do my prophets no harm." At first glance, these words may seem straightforward, but their implications are profound and complex. When the text warns against touching the anointed, it goes beyond physical contact. It reminds us to treat all individuals consecrated by God with reverence and respect. This includes priests, kings, and ordinary believers chosen and set apart for a specific purpose.

. . .

Furthermore, the command to not harm the prophets extends to a broader principle of not obstructing or undermining those who carry out God's work. In the biblical context, prophets are not just fortune-tellers but messengers of divine truth. By harming them, we risk silencing God's voice and disrupting the flow of His message to humanity.

Tragically, history is filled with examples of prophets being persecuted, silenced, or even killed for speaking truth to power. The admonition in Psalm 105:15 serves as a cautionary reminder of the consequences of such actions and underscores the importance of upholding and protecting those who convey God's message.

While the text calls for respect and protection of the anointed and prophets, it does not absolve them of accountability. When leaders or spiritual figures abuse authority or cause harm, the community must address these issues with wisdom and discernment.

Seeking resolution and reconciliation, as outlined in Matthew 18:15-16, is crucial in addressing conflicts within the faith community. By following the principles of accountability and reconciliation, individuals can uphold the values of justice and integrity while preserving the unity of believers.

In personal experiences within the church, the support and guidance of spiritual leaders play a vital role in navigating challenges and seeking redress for wrongs committed. Conflicts can be resolved and relationships restored through open communication, mutual respect, and adherence to biblical principles.

. . .

REMEMBER, THE COMMAND TO "NOT HARM" extends to physical actions and our words and attitudes toward one another. By embracing a spirit of love, understanding, and forgiveness, we can honor the sanctity of God's anointed ones and prophets while upholding the values of justice and righteousness in our communities.

10

FIELD MEDICS ON THE BATTLEFIELD

You will undergo basic combat training for 10 weeks to ensure you are physically and mentally prepared to serve in the Army. After completing this training, you'll proceed to advanced individual training (AIT), which can last anywhere from 16 to 68 weeks, depending on your specialization and the area of medicine you will study. It's important to note that this training prepares you for frontline duty.

When my husband and I first met, he worked as a 41K MOS, repairing and rebuilding reproduction equipment such as offset presses, plate graining machines, and coating machines. After twelve years, he pursued a new career path and became a 68W, working as a field medic. At that time, we were actively involved in ministry, and I was intrigued by his unique skill set. As people of faith, we understood the critical role of prophets in warning God's people and helping them deal with trauma in their cultures. However, I noticed a significant change in my husband when he was called to his third and fourth tours in Iraq.

. . .

In the middle of the night, he called me and shared that he had been assigned to a mission as a medical aid to rescue someone in distress. During the mission, he recounted how their truck was hit by an IED, resulting in critical injuries to one of their comrades. He explained that the injured person was in such dire condition that he had to administer a shot of morphine to stabilize them until they could be taken to Landstuhl, Germany.

I asked all **the questions that wives typically do:**

- "Are you okay?"
- "Are you hurt?"
- "Were you scared?"
- "How did you assist while you were afraid?"

Once he returned home, he shared his experience of how his training helped him in that critical situation and explained the necessity of the morphine injection. He elaborated that, in cases where they are called to tend to a soldier injured by an IED, they are instructed to administer a generous portion of painkillers to manage the pain. This helps keep the soldier comfortable until they can be transported to the hospital for further treatment. Essentially, the field medic's job is to ensure that the soldier is as stable as possible until the transport process is completed and they receive proper medical attention.

During these moments of reflection, I truly understood the depth of my husband's commitment to his role as a field medic. The sacrifices he made, the risks he faced, and the emotional toll it took on him became more apparent with each story he shared. He was heroic and dedicated to saving lives and providing critical care in challenging circumstances.

As his partner, I grappled with a mix of emotions—pride in his bravery, concern for his well-being, and profound respect for his work. It was a journey of learning and growth for both of us as we navigated the complexities of his experiences on the battlefield and how they impacted our lives.

Through it all, one thing remained constant: his unwavering commitment to his duty as a field medic. The training he had received, the skills he had honed, and the instincts he had developed were all tested in those critical crises. Time and time again, he rose to the occasion, demonstrating competence and compassion in the face of adversity.

Each mission, rescue, and life saved was a testament to his courage and calling. As I stood by his side, offering support and understanding, I realized that being a field medic was not just a job for him—it was a calling, a vocation that defined him in ways that words could never fully capture.

In March 2007, I received a call that every wife dreads: I was informed that something had happened to my husband. I was quickly connected to a satellite phone and could hear his voice on the other end. He sounded energetic, clear, and lively, as if nothing was wrong. However, he then delivered the devastating news that he had suffered a severed spine. The call abruptly disconnected, and I was told to pack my bags; someone would pick me up in the morning.

Healing Is On The Battlefield

As we conclude this book, I want to encourage readers to take the time to learn from their adversities. It's hard to think clearly, eat properly, and sometimes trust God during difficult seasons. Your emotions may conflict with reality, and your mental state feels like it's walking a tightrope of sanity. I know what it's like to be under the pressure of losing my mind, especially after being blocked by those in authority who sought to cover their sins. They not only used me as a scapegoat but also took pleasure in ensuring that, for the next ten years, they could destroy relationships once dear to me and close doors to once-open opportunities.

LIKE MY HUSBAND, I felt called to help others, teaching and speaking like a prophet. Yet, in my time of need, I felt abandoned. Isn't it ironic how someone can harm you emotionally or spiritually and then turn around and accuse you of wrongdoing as if they are innocent? You hope for justice, but often the truth remains hidden. You may seek healing amid this conflict, yet it feels elusive.

THIS BOOK WAS NOT WRITTEN out of bitterness or unforgiveness but to share what I learned and what God wanted me to understand. What truly harms you is often visible, but what you can't see can be even more damaging. Like many prophets, I retreated into the cave of obscurity, blaming God instead of trusting His plan. In hindsight, I was too immature to understand God's purpose for me, but He was preparing me for the journey ahead.

OVER THE YEARS, God revealed that my calling was similar to Paul's, where rejection and persecution would be part of my experience. However, through these trials, God would unveil His wisdom and truth to generations of prophets who walk in the authority He has assigned.

Being in that cave gave me a sense of healing that helped me maintain my sanity. Like Elijah, I learned to trust in God's provision for my needs and cherished the moments spent connecting with those who nourished my spirit. The Holy Spirit truly is the ultimate healing agent!

Rev. 10:10-11;

10 And I took the little book out of the angel's hand, and ate it up, and it was in my mouth sweet as honey: as soon as I had eaten it, my belly was bitter. And he said unto me, Thou must prophesy again before many peoples, nations, tongues, and kings.

During the seven years I received divine revelations, I found solace and enlightenment in the solitude of a cave, reminiscent of the spiritual retreats of biblical figures like John and Paul. This period of seclusion was a time of deep reflection and a transformative experience that allowed me to explore profound teachings and messages in preparation for my sacred duty to minister to the marginalized and vulnerable in society.

Much like the prophets of old who sought divine guidance in isolated places, the cave became a sanctuary for my personal growth and spiritual journey. Within the quiet confines of this natural refuge, I discovered the power of introspection, realizing that true healing comes from accepting oneself and releasing the burdens that weigh heavily on the soul.

. . .

JUST AS CHRIST encouraged Peter to share his wisdom with others, I have embraced my calling to extend a helping hand to those in need. Through my experiences and revelations, I understand the importance of reaching out to struggling people and offering them hope, compassion, and healing.

AS I EMBARK on my journey, I recognize profound parallels between my path and the timeless quests for spiritual enlightenment depicted in the Bible. Just as the stories of old have shaped the faith of many, my experiences have transformed me into a humble and empathetic servant. This transformation has instilled in me a deep commitment to bringing light and solace to the marginalized and downtrodden individuals I encounter. My mission is clear: to offer a helping hand and empower others to rise above their challenges. Through teaching and sharing knowledge, I aim to equip those around me with the tools they need to navigate the battles of life with courage and resilience. Thus, no prophet is left behind!

"But I have prayed for thee, that thy faith fail not: and when thou art converted, strengthen thy brethren" (Luke 22:31–32).

NOTES

4. Why Identity Matters

1. Thompson J, Lockhart W, Roach W, Atuel H, Bélanger S, Black T, et al. *Veterans' identities and well-being in transition to civilian life—A resource for Policy Analysts, Program Designers, Service Providers and Researchers.* 2017

Made in the USA
Monee, IL
05 November 2026

67377157R00069